The Sewing Machine:

Its Invention and Development

The Sewing Machine
Its Invention and Development

GRACE ROGERS COOPER

PUBLISHED BY THE SMITHSONIAN INSTITUTION PRESS FOR THE

National Museum of History and Technology

Smithsonian Institution Washington, D.C. 1976

All rights reserved
First edition, 1968
Second printing, 1969
Second edition, revised and expanded, 1976

Printed in the United States of America

Library of Congress Cataloging in Publication Data
Cooper, Grace Rogers.
   The sewing machine: its invention and development
   First published in 1968 under title: The invention
     of the sewing machine
   Bibliography: p.
   Includes indexes.
   1. Sewing machines—History.   I. Title.
TJ1507.06   681'.7677   75-619415

ISBN 0-87474-330-3

Contents

Preface	vii
Acknowledgments	ix
1. Early Efforts	1
2. Elements of a Successful Machine	17
3. The "Sewing-Machine Combination"	39
4. Less-Expensive Machines	43
Appendixes	
I. Notes on the Development and Commercial Use of the Sewing Machine	57
II. American Sewing-Machine Companies of the 19th Century	65
III. Sewing-Machine Styles	158
IV. 20th-Century Sewing Machines	164
V. Chronological List of U.S. Sewing-Machine Patent Models in the Smithsonian Collections	206
VI. 19th-Century Sewing-Machine Leaflets in the Smithsonian Collections	215
VII. A Brief History of Cotton Thread	216
VIII. Biographical Sketches	217
Bibliography	224
Indexes	
Geographical Index to Companies Listed in Appendix II	229
Alphabetical Index to Patentees Listed in Appendix V	232
General Index to Chapters 1-4	237

Preface

This is a new and expanded edition of *The Invention of the Sewing Machine,* first published in 1968 and reprinted in 1969. The response to that first effort on the subject brought forth many good questions about late 19th- and early 20th-century machines. We now wish to share additional data on these machines of the most colorful period of sewing-machine history.

The American-made sewing machine, like the horse and buggy, is past. A little more than one hundred years ago American inventors created and American manufacturers supplied this first widely advertised consumer product. Today, economic competition has forced the U.S.-manufactured sewing machine out of the market. The American companies that have survived have done so by relying on foreign manufacturing plants. Thus, an era has ended.

From the beginning the sewing machine had wide appeal. It had no instrument panel with push-button controls. It was not operated electronically or jet-propelled; but, to the masses of 19th-century people the sewing machine was as awe-inspiring as the space capsule to their 20th-century descendants. Appearing in elaborately encased displays at international expositions, the sewing machine was the mechanical wonder of the century to those who saw and used it. Expensive, yes, but considering the work it would do—a machine that could sew—and the time it could save, the cost was more than justified. The sewing machine pioneered installment buying and patent pooling. It also weathered the protests of the seamstresses and tailors who feared this new machine was a threat to their means of earning a living.

The practical sewing machine is not the result of one man's genius, but rather the culmination of a century of thought, work, trials, failures, and partial successes of a long list of inventors. History is too quick to credit one or two men for an important invention and to forget the work that preceded and prodded each man. It is no discredit to Elias Howe to state that he *did not invent the sewing machine.* Howe's work with the sewing machine was important, and he did patent certain improvements, but his work was but one step along the way. It is for the reader to decide whether it was the turning point. Of equal importance to the story of the invention is the history of the sewing-machine's manufacture.

The relatively few companies of the first decade, the 1850s, blossomed into well over one hundred after the expiration of the major patents and the dissolution of the Sewing-Machine Combination in 1877. A catalog list of more than one hundred and fifty of these 19th-century companies is included in this study. Still, the list is incomplete. Many of the companies remained in business a very short time or kept their activities a secret to avoid payment of royalties to patent holders. Evidence of these companies may be extant, but it lies submerged in local histories or family records. Some new information came to light in response to the first edition of this history; it is hoped that this edition will surface even more. Many of the companies and their machines bore the same name. Although new styles were introduced and the machines were improved, the name remained the same reflecting the strength of the company and the product.

The companies of the late 19th century, who did not have the first twenty or thirty years to back their product's name, introduced an ever-increasing number of colorful names—American Beauty, Bastedor O'Damode, Cherokee, Full Dinner Pail, Golden Rule, and thousands more. Even when there were fewer than twenty-five companies in active production in the early twentieth century, one company alone produced sewing machines sold under hundreds of different names. In general, the machines were not very different from each other except for the name. Style characteristics of the company usually enables one to identify the manufacturer; many of these machines were sold through distributors with their choice of name. Over three thousand of these names are listed in this edition. As in furniture, sewing-machine tables and cabinets changed too, dutifully reflecting the era, new manufacturing techniques, and fickle fashion—in cabinet detail, ironwork, and mechanical innovation.

Old sewing machines are collectible. The machines were well-built, mechanically precise, durable, and beautifully finished. With reasonable care they still look good after seventy-five or a hundred years. With not-so-good care, they can be put into good condition with mechanical skill, patience, and careful refinishing. There are varying reasons to collect these machines. For a museum, it might be the history of a company or industry, the significance of the invention, the effect on the developing ready-made clothing industry, as a popular 19th-century household appliance, or the far-reaching effect on all types of manufactures of stitched textiles. For the private collector, it might be the appeal of the small hand-turned machines that can be attractively shelved as mechanical curiosities, the decorative appeal of the ironwork or the beauty of the woodwork of the cabinet machines. For the individual, it might be a prized family heirloom or the appeal of a single, isolated, eye-catching example.

The current antique value of the sewing machines is not included in this study. Museum staffs do not make monetary appraisals. The prices quoted in this work are the prices as advertised when the machines were new.

Acknowledgments

As stated in the first edition of this book, I am greatly indebted to the late Dr. Frederick Lewton, whose interest in the history of the sewing machine initiated the collecting of information by the Smithsonian Institution's Division of Textiles; to the Singer Manufacturing Company for their gift of an excellent collection of 19th-century sewing machines; and to The Henry Ford Museum and Greenfield Village for having permitted me to study their collection.

For this edition, I would also like to acknowledge the supportive efforts of the entire staff of the Division of Textiles, but most especially to Julia Fiechter and Cynthia Cunningham for their tireless hours proofing thousands of sewing-machine names. I also wish to express my appreciation for the clerical assistance I received from the Department of Applied Arts; the photographic work of the Smithsonian Office of Printing and Photographic Services; the watchful eyes of a sympathetic editor, Louise Heskett; and for an energetic summer volunteer, Anne Hoke, who wrestled with heavy sewing machines and juggled an endless number of cards.

I would also like to acknowledge the many hundreds of interested owners and collectors of sewing machines and sewing-machine history both in the United States and abroad who have written to me for and with information about their sewing machines since the first printing of *The Invention of the Sewing Machine* in 1968. It has been this never-ending tide of interest that encouraged me to expand this work.

Grace Rogers Cooper
Curator of Textiles

Washington, D.C.
March 1976

Chapter One

Figure 1.—After almost a century of attempts to invent a machine that would sew, the practical sewing machine evolved in the mid-19th century. This elegant, carpeted salesroom of the 1870s, with fashionable ladies and gentlemen scanning the latest model sewing machines, reflects the pinnacle reached by the new industry in just a few decades. This example, one of many of its type, is the Wheeler and Wilson sewing-machine offices and salesroom, No. 44 Fourteenth Street, Union Square, New York City. From *The Daily Graphic*, New York City, December 29, 1874. (Smithsonian photo 48091–A.)

Early Efforts

To 1800

For thousands of years, the only means of stitching two pieces of fabric together had been with a common needle and a length of thread. The thread might be of silk, flax, wool, sinew, or other fibrous material. The needle, whether of bone, silver, bronze, steel, or some other metal, was always the same in design—a thin shaft with a point at one end and a hole or eye for receiving the thread at the other end. Simple as it was, the common needle (fig. 2) with its thread-carrying eye had been an ingenious improvement over the sharp bone, stick, or other object used to pierce a hole through which a lacing then had to be passed.[1] In addition to utilitarian stitching for such things as the making of garments and household furnishings, the needle was also used for decorative stitching, commonly called embroidery. And it was for this purpose that the needle, the seemingly perfect tool that defied improvement, was first altered for ease of stitching and to increase production.

One of the forms that the needle took in the process of adaptation was that of the fine steel hook. Called an *aguja* in Spain, the hook was used in making a type of lace known as *punto de aguja*. During the 17th century after the introduction of chainstitch embroideries from India, this hook was used to produce chainstitch designs on a net ground.[2] The stitch and the fine hook to make it were especially adaptable to this work. By the 18th century the hook had been reduced to needle size and inserted into a handle, and was used to chainstitch-embroider woven fabrics.[3] In France the hook was called a crochet and was sharpened to a point for easy entry into the fabric (fig. 3). For stitching, the fabric was held taut on a drum-shaped frame. The hooked needle pierced the fabric, caught the thread from below the surface and pulled a loop to the top. The needle reentered the fabric a stitch-length from the first entry and caught the thread again, pulling a second loop through the first to which it became enchained. This method of embroidery permitted for the first time the use of a continuous length of thread. At this time the chainstitch was used exclusively for decorative embroidery, and from the French name for drum—the shape of the frame that held the fabric—the worked fabric came to be called tambour embroidery. The crochet[4]

---

[1] Charles M. Karch, *Needles: Historical and Descriptive* (12 Census U.S., vol. X, 1902), pp. 429–432.

[2] Florence Lewis May, *Hispanic Lace and Lace Making* (New York, 1939), pp. 267–271.

[3] Diderot's *L'Encyclopédie, ou dictionnaire raisonné des sciences, des arts et des métiers . . .* , vol. II (1763), Plates Brodeur, plate II.

[4] The term "crochet," as used today, became the modern counterpart of the Spanish *punto de aguja* about the second quarter of the 19th century.

or small hooked needle soon became known as a tambour needle.

In 1755 a new type of needle was invented for producing embroidery stitches. This needle had to pass completely through the fabric two times (a through-and-through motion) for every stitch. The inventor was Charles F. Weisenthal, a German mechanic living in London who was granted British patent 701 for a two-pointed needle (fig. 4). The invention was described in the patent as follows:

> The muslin, being put into a frame, is to be worked with a needle that has two points, one at the head, and the other point as a common needle, which is to be worked by holding it with the fingers in the middle, so as not to require turning.

It might be argued that Weisenthal had invented the eye-pointed needle, since he was the first inventor to put a point at the end of the needle having the eye. But, since his specifically stated use required the needle to have two points and to be passed completely through the fabric, Weisenthal had no intention of utilizing the very important advantage that the eye-pointed needle provided, that of *not* requiring the passage of the needle through the fabric as in hand sewing.

While no records can be found to establish that Weisenthal's patent was put to any commercial use during the inventor's lifetime, the two-pointed needle with eye at midpoint appeared in several 19th-century sewing-machine inventions.

The earliest of the known mechanical sewing devices produced a chain or tambour stitch, but by an entirely different principle than that used with either needle just described. Although the idea was incorporated into a patent, the machine was entirely overlooked for almost a century as the patent itself was classed under wearing apparel. It was entitled "An Entire New Method of Making and Completing Shoes, Boots, Splatterdashes, Clogs, and Other Articles, by Means of Tools and Machines also Invented by Me for that Purpose, and of Certain Compositions of the Nature of Japan or Varnish, which will be very advantageous in many useful Applications." This portentously titled British patent 1,764 was issued to an English cabinetmaker, Thomas Saint, on July 17, 1790. Along with accounts of several processes for making various varnish compositions, the patent contains descriptions of three separate machines; the second of these was for "stitching, quilting, or sewing." Though far from practical, the machine incorporated

Figure 2.—PRIMITIVE NEEDLE. Bronze. Egyptian (Roman period, 30 B.C.–A.D. 642). (Smithsonian photo 1379–A.)

several features common to a modern sewing machine. It had a horizontal cloth plate or table, an overhanging arm carrying a straight needle, and a continuous supply of thread from a spool. The motion was derived from the rotation of a hand crank on a shaft, which activated cams that produced all the actions of the machine.

One cam operated the forked needle (fig. 5) that pushed the thread through a hole made by a preceding thrust of the awl. The thread was caught by a looper and detained so that it then became enchained in the next loop of thread. The patent described thread tighteners above and below the work and an adjustment to vary the stitches for different kinds of material. Other than the British patent records, no contemporary reference to Saint's machine has ever been found. The stitching-machine contents of this patent was happened on by accident in 1873.[5] Using the patent description, a Newton Wilson of London attempted to build a model of Saint's machine in 1874.[6] Wilson found, however, that it was necessary to modify the construction before the machine would stitch at all.

---

[5] *Sewing Machine News* (1880), vol. 1, no. 7, p. 2.
[6] This model of Saint's machine was bequeathed by Mr. Wilson to the South Kensington Museum, London, England.

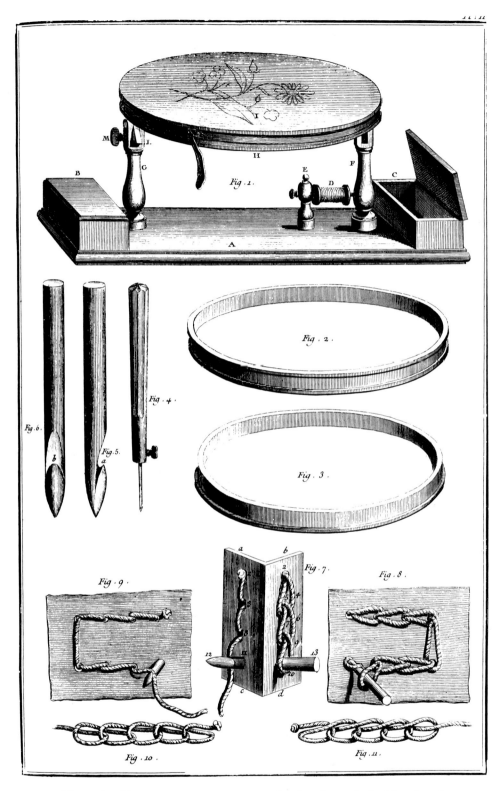

Figure 3.—Tambour needle and frame, showing the method of forming the chainstitch, from the Diderot Encyclopedia of 1763, vol. II, *Plates Brodeur*, plate II. (Smithsonian photo 43995–C.)

This raised the question whether Saint had built even one machine. Nevertheless, the germ of an idea was there, and had the inventor followed through the sewing machine might have been classed an 18th-century rather than a 19th-century contribution.

1800–1820

There is no doubt that the successful late-18th-century improvements in spinning and weaving methods, resulting in increased production of fabrics, had a great effect in spurring inventors to ideas of stitching by machinery. Several efforts were made during the first two decades of the 19th century to produce such machines.

On February 14, 1804, a French patent was issued to Thomas Stone and James Henderson for a "new mechanical principle designed to replace handwork in joining the edges of all kinds of flexible material, and particularly applicable to the manufacture of clothing." [7] The machine used a common needle and made an overcast stitch in the same manner as hand sewing. A pair of jaws or pincers, imitating the action of the fingers, alternately seized and released the needle on each side of the fabric. The pincers were attached to a pair of arms arranged to be moved backward and forward by "any suitable mechanism." [8] This machine was capable of making curved or angular as well as straight seams, but it was limited to carrying a short length of thread, necessitating frequent rethreading. The machine may have had some limited use, but it was not commercially successful.

On May 30 of the same year John Duncan, a Glasgow manufacturer, was granted British patent 2,769 for "a new and improved method of tambouring, or raising flowers, figures or other ornaments upon muslins, lawns and other cottons, cloths, or stuffs." This machine made the chainstitch, using not one but many hooked needles that operated simultaneously. The needles, attached to a bar or carrier, were pushed through the vertically held fabric from the upper right side, which in this case was also the outer side. After passing through it, they were supplied with thread from spools by means of peculiarly formed hooks or thread carriers. The thread was twisted around the needle above the hook, so as to be caught by it, and drawn through to the outer surface. The shaft of the needle was grooved on the hook side and fitted with a slider. This slider closed upon the retraction of the needle from the fabric, holding the thread in place and preventing the hook from catching. The fabric was stretched between two rollers set in an upright frame capable of sliding vertically in a second frame arranged to have longitudinal motion. The combination of these two motions was sufficient to produce any required design. The principle developed by Duncan was used on embroidery machines, in a modified form, for many years. Of several early attempts, his was the first to realize any form of success.

A type of rope-stitching machine, which might be

Figure 4.—WEISENTHAL's two-pointed needle, 1755.

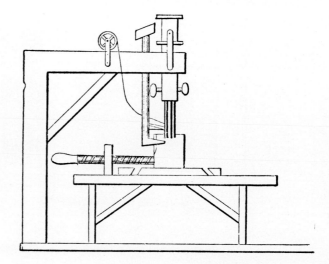

Figure 5.—SAINT'S SEWING MACHINE, 1790.
(Smithsonian photo 42490–A.)

---
[7] *Sewing Machine News* (1880), vol. 1, no. 8, p. 2.
[8] Ibid.

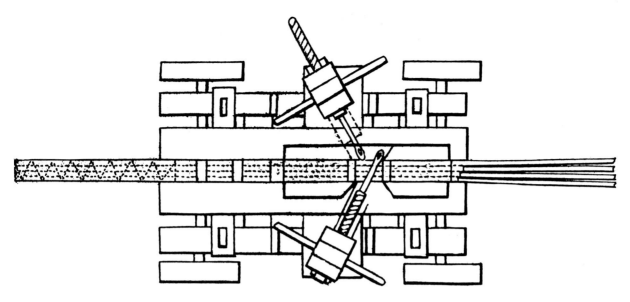

Figure 6.—CHAPMAN'S SEWING MACHINE, first eye-pointed needle, 1807. (Smithsonian photo 33299–K.)

considered unimportant to this study, must be included because of its use of the eye-pointed needle, the needle that was to play a most important part in the later development of a practical sewing machine. The earliest reference to the use of a needle with an eye not being required to be passed completely through the fabric it was stitching is found in a machine invented by Edward Walter Chapman, for which he and William Chapman were granted British patent 3,078 on October 30, 1807. The machine (fig. 6) was designed to construct belting or flat banding by stitching together several strands of rope that had been laid side by side. Two needles were required and used alternately. One needle was threaded and then forced through the ropes. On the opposite side the thread was removed from the eye of the first needle before it was withdrawn. The second needle was threaded and the operation repeated. The needles could also be used to draw the thread, rather than push it, through the ropes with the same result. While being stitched, the ropes were held fast and the sewing frame and supporting carriage were moved manually as each stitch was made. Such a machine would be applicable only to the work described, since the necessity of rethreading at every stitch would make it impractical for any other type of sewing.

Another early machine reported to have used the eye-pointed needle to form the chainstitch was invented about 1810 by Balthasar Krems,[9] a hosiery worker of Mayen, Germany. One knitted article produced there was a peaked cap, and Krems' machine was devised to stitch the turned edges of the cap,[10] which was suspended from wire pins on a moving wheel. The needle of the machine was attached to a horizontal shaft and carried the thread through the fabric. The loop of thread was retained by a hook-shaped pin to become enchained with the next loop at the reentry of the needle. Local history reports that this device may have been used as early as 1800, but the inventor did not patent his machine and apparently made no attempt to commercialize it. No contemporary references to the

---

[9] ERICH LUTH, *Ein Mayener Strumpfwirker, Balthasar Krems, 1760–1813, Erfinder der Nähmaschine*, p. 10, states that the machine used an eye-pointed needle. WILHELM RENTERS, *Praktisches wissen von der Nähmaschine*, p. 4, states that Krems used a hooked needle. Renters probably mistook the hooked retaining pin for the needle.

[10] Dr. Dahmen, Burgermeister of Mayen, stated in a letter of October 8, 1963, that the original Krems machine was turned over to the officials of Mayen by Krems' descendants about the turn of the century. He verified that the machine used an eye-pointed needle. About 1920 the machine was placed in the Eifelmuseum in Genovevaburg; some of the unessential parts were restored. The machine now at this museum is the one pictured in Luth's book. A replica of the machine is in the Deutsches Museum, Munich, Germany.

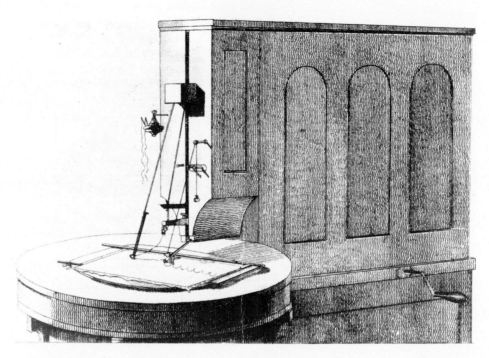

Figure 7.—MADERSPERGER'S 1814 SEWING MACHINE. Illustration from a pamphlet by the inventor entitled *Beschreibung einer Nähmaschine*, Vienna, ca. 1816. (Smithsonian photo 49373.)

machine could be found, and use of the machine may have died with the inventor in 1813.

About the same time, Josef Madersperger, a tailor in Vienna, Austria, invented a sewing machine, which was illustrated (fig. 7) and described in a 15-page pamphlet published about 1816.[11] On May 12, 1817, a Vienna newspaper wrote of the Madersperger machine: "The approbation which his machine received everywhere has induced his Royal Imperial Majesty, in the year 1814, to give to the inventor an exclusive privilege [patent] which has already been mentioned before in these papers."[12] Madersperger's 1814 machine stitched straight or curving lines. His second machine stitched small semicircles, as shown in the illustration, and also small circles, egg-shaped figures, and angles of various degrees. The machine, acclaimed by the art experts, must therefore have been intended for embroidery stitching. From the contemporary descriptions and the illustration, the machine is judged to have made a couched stitch—one thread was laid on the surface of the fabric and stitched in place with a short thread carried by a two-pointed needle of the type invented by Weisenthal. Two fabrics could have been stitched together, but not in the manner required for tailoring. The machine must have had many deficiencies in the tension adjustment, feed, and related mechanical operations, for despite the published wishes for success the inventor did not put the machine into practical

---

[11] JOSEF MADERSPERGER, *Beschreibung einer Nähmaschine* (Vienna, ca. 1816). The exact date of this small booklet is not known. In the booklet Madersperger reports that he had received a patent in 1814 for his *first* machine adapted to straight sewing. However, the machine described and illustrated in this booklet was one that could stitch semicircles and small figures. In *Kunst und Gewerbeblatt*, a periodical (Munich, Germany, 1817, pp. 336–338), reference is made to the Madersperger machine and a statement to the effect that the inventor had published a leaflet describing his machine. The leaflet referred to is believed to be the one under discussion. For this reason it must have been published between 1814 and 1817, therefore ca. 1816. The only copy of this booklet known to this author is in the New York Public Library. It was probably not known to authors Luth and Renters. The author wishes to thank Miss Rita J. Adrosko of her staff for her important help in translating these German publications.

[12] *Sewing Machine Times* (1907), vol. 26, no. 865, p. 1.

not mentioned in the earlier edition. The writer of the article on sewing machines states that John Knowles invented and constructed a sewing machine, which used a single thread and a two-pointed needle with the eye in the middle to form the backstitch. This information must have come to light after the first edition was published, but from where and by whom is not known. Other sources state that two men, Adams and Dodge, produced this machine in Monkton.[14] While still others credit the Reverend John Adam Dodge, assisted by a mechanic by the name of John Knowles, with the same invention in the same location.[15] Vermont historical societies have been unable to identify the men named or to verify the story of the invention.[16] The importance of the credibility of this story, if proved, rests in the fact that it represents the first effort in the United States to produce a mechanical stitching device.

1820–1845

American records of this period are incomplete as a result of the Patent Office fire of 1836, in which most of the specific descriptions of patents issued to that date were destroyed. Patentees were asked to provide another description of their patents so that these might be copied, but comparatively few responded and only a small percentage was restored. Thus, although the printed index of patents [17] lists Henry Lye as patenting a machine for "sewing leather, and so forth" on March 10, 1826, no description of the machine has ever been located. Many patents whose original claim was for only a mechanical awl to pierce holes in leather or a clamp to hold leather for hand stitching were claimed as sewing devices once a practical machine had evolved. But no evidence has ever been found that any of these machines performed the actual stitching operation.

The first man known to have put a mechanical

Figure 8.—An engraving of Thimonnier and his sewing machine of 1830, from *Sewing Machine News*, 1880. (Smithsonian photo 10569–C.)

operation.[13] Years later Madersperger again attempted to invent a sewing machine using a different stitch (see p. 13).

A story persists that about 1818–1819 a machine that formed a backstitch, identical to the one used in hand sewing, was invented in Monkton, Vermont. The earliest record of this machine that this author has found was in the second or 1867 edition of *Eighty Years of Progress of the United States;* the machine is

---

[13] There are no known models of these early Madersperger machines in existence. Although the *Sewing Machine Times* reported in the 1907 issue that the 1814 sewing machine was then on exhibition in the Museum of the Vienna Polytechnic, the illustration shown was of Madersperger's 1839 machine. In a letter from the director of the Technisches Museum für Industrie und Gewerbe in Vienna, received in 1962, it was stated that the original 1814 Madersperger machine was in their museum. The photographs that were sent, however, were of the 1839 machine. This machine is entirely different from the 1814–1817 machine, as can readily be seen by the reader (figs. 7 and 10).

[14] John P. Stambaugh, *A History of the Sewing Machine* (Hartford, Conn., 1872), p. 13; *Sewing Machine News* (July 1880), vol. 1, no. 12, p. 4.

[15] "Sewing Machines," *Johnson's Universal Cyclopaedia* (New York, 1878), vol. 4, p. 205. The 1874 edition does not include this reference to the Reverend John Adam Dodge.

[16] Letters to the author from the Vermont Historical Society (Nov. 13, 1953) and the Bennington Historical Museum and Art Gallery (May 2, 1953).

[17] Edmund Burke. Commissioner of Patents, *List of Patents for Inventions and Designs Issued by the United States from 1790 to 1847* (Washington, 1847).

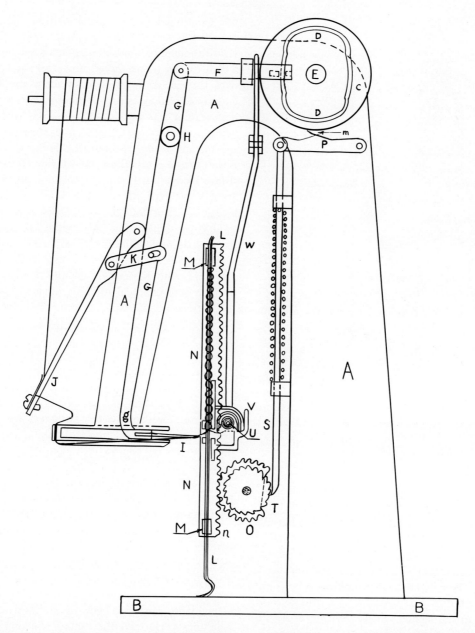

Figure 9.—An adapted drawing of Hunt's sewing machine published by the *Sewing Machine News*, vol. 2, no. 8, 1881, to give some idea of its construction and operation. "The frame of the machine (A) rested on a base (B) that was supported by a table. The wheel (C) worked on a central shaft (E) and was set in motion by hand or foot power. On the front of the wheel (C) was a raised cam (D) into which the connecting rod (F) engaged to communicate motion to the vibrating arm (G) pivoted to the frame at (H) and carrying at the end (g) the curved needle (I). The take-up (J) served to tighten the thread after each stitch; it was connected to the vibrating arm by a rod (K). The cloth (L) was held in a vertical position between the fingers or nippers (M), which were attached to the frame. The bar (N) was toothed on one side (n) to mesh with the geared wheel (o). The lever (P) was operated by a cam (m) upon the periphery of the wheel (C), and carried the vertical pawl (S) which meshed with the ratchet (T) and moved the cloth as each stitch was made. The shuttle (U) worked in its race (V); it was operated by the vibrating lever (W), the upper end of which engaged into a groove on the face of the wheel (C)." (Smithsonian photo 42554.)

ewing device into commercial operation was Barthelemy Thimonnier,[18] a French tailor. After several years of fruitless effort he invented a machine for which he received a French patent in 1830.[19] The machine (fig. 8) made a chainstitch by means of a barbed or hooked needle. The vertically held needle worked from an overhanging arm. The needle thrust through the fabric laid on the horizontal table, caught a thread from the thread carrier and looper beneath the table, and brought a loop to the surface of the fabric. When the process was repeated the second loop became enchained in the first. The needle was moved downward by the depression of a cord-connected foot treadle and was raised by the action of a spring. The fabric was fed through the stitching mechanism manually, and a regular rate of speed had to be maintained by the operator in order to produce stitches of equal length. A type of retractable thimble or presser foot was used to hold the fabric down as required.

The needle, and the entire machine, was basically an attempt to mechanize tambour embroidery, with which the inventor was quite familiar. Although this work, which served as the machine's inspiration, was always used for decorative embroidery, Thimonnier saw the possibilities of using the stitch for utilitarian purposes. By 1841 he had 80 machines stitching army clothing in a Paris shop. But a mob of tailors, fearing that the invention would rob them of a livelihood, broke into the shop and destroyed the machines. Thimonnier fled Paris, penniless. Four years later he had obtained new financial help, improved his machine to produce 200 stitches a minute, and organized the first French sewing-machine company.[20] The Revolution of 1848, however, brought this enterprise to an early end. Before new support could be found other inventors had appeared with better machines, and Thimonnier's was passed by. In addition to the two French patents, Thimonnier also received a British patent with his associate Jean Marie Magnin in 1848 and one in the United States in 1850. He achieved no financial gain from either of these and died a poor man.

While Thimonnier was developing his chainstitch machine in France, Walter Hunt,[21] perhaps best described as a Yankee mechanical genius, was working on a different kind of sewing machine in the United States. Sometime between 1832 and 1834 he produced at his shop in New York a machine that made a lockstitch.[22] This stitch was the direct result of the mechanical method devised to produce the stitching and represented the first occasion an inventor had not attempted to reproduce a hand stitch. The lockstitch required two threads, one passing through a loop in the the other and both interlocking in the heart of the seam. At the time Hunt did not consider the sewing machine any more promising than several other inventions that he had in mind, and, after demonstrating that the machine would sew, he sold his interest in it for a small sum and did not bother to patent it.

A description—one of few ever published—and sketch of a rebuilt Hunt machine (fig. 9) appeared in an article in the *Sewing Machine News* in 1881.[23] The important element in the Hunt invention was an eye-pointed needle working in combination with a shuttle carrying a second thread. Future inventors were thus no longer hampered by the erroneous idea that the sewing machine must imitate the human hands and fingers. Though Hunt's machine stitched short, straight seams with speed and accuracy, it could not sew curved or angular work. Its stitching was not continuous, but had to be reset at the end of a short run. The validity of Hunt's claim as the inventor of the lockstitch and the prescribed method of making it was argued many times, especially during the Elias Howe patent suits of the 1850s. The decision against Hunt was not a question of invention,[24] but one of right to ownership or control. Hunt did little to promote his sewing machine and sold it together with the right to patent to George A. Arrowsmith.

---

[18] See biographical sketch, pp. 217–218.

[19] French patent issued to Barthelemy Thimonnier and M. Ferrand (who was a tutor at l'Ecole des Mines, Saint-Etienne, and helped finance the patent), July 17, 1830.

[20] The company was located at Villefranche-sur-Saône, but no name is recorded. See J. Granger, *Thimonnier et la machine à coudre* (1943), p. 16.

---

[21] See biographical sketch, p. 218.

[22] The earliest known reference in print to Walter Hunt's sewing machine is in *Sewing by Machinery: An Exposition of the History of Patentees of Various Sewing Machines and of the Rights of the Public* (I. M. Singer & Co., 1853). A more detailed story of Hunt's invention is in *Sewing Machine News* (1880–81), vol. 2, no. 2, p. 4; no. 4, p. 5; and no. 8, pp. 3 and 8.

[23] Vol. 2, no. 8, p. 3.

[24] In the opinion and decision of C. Mason, Commissioner of the Patent Office, offered on May 24, 1854, for the Hunt vs. Howe interference suit, Mason stated: "He [Hunt] proves that in 1834 or 1835 he contrived a machine by which he actually effected his purpose of sewing cloth with considerable success."

Figure 10.—MADERSPERGER's 1839 sewing machine. Madersperger's machine consisted of two major parts: the frame, which held the material, and the stitching mechanism, called the hand. The hand shown here is an original model. (*Photo courtesy of Technisches Museum für Industrie und Gewerbe, Vienna.*)

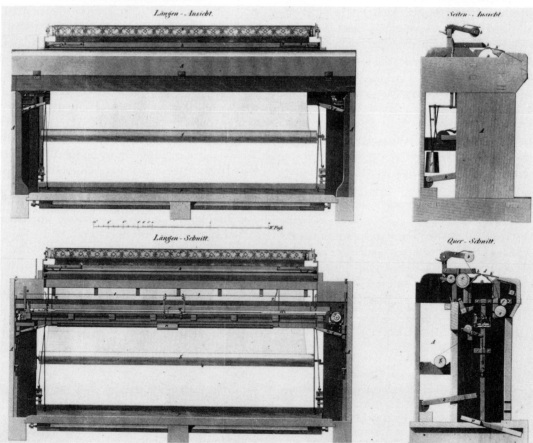

For over fifteen years, from the mid-1830s to the early 1850s, the machine dropped out of sight. When the sewing-machine litigation developed in the 1850s, the I. M. Singer company searched out the Hunt machine, had the inventor rebuild one,[25] and attempted to use this to break the Howe patent. The plan did not work. The Honorable Charles Mason, Patent Commissioner, reported:

> When the first inventor allows his discovery to slumber for eighteen years, with no probability of its ever being brought into useful activity, and when it is only resurrected to supplant and strangle an invention which has been given to the public, and which has been made practically useful, all reasonable presumption should be in favor of the inventor who has been the means of conferring the real benefit upon the world.[26]

Hunt's machine was an invention of the 1830s, but only because of the patent litigation was it ever heard of again.

During the time that a potentially successful sewing machine was being invented and forgotten in America, Josef Madersperger of Austria made a second attempt to solve the mechanical stitching problem. In 1839 he received a second patent on a machine entirely different from his 1814 effort. It was similar to Hunt's in that it used an eye-pointed needle and passed a thread through the loop of the needle-thread—the thread carried by the needle—to lock the stitch. Madersperger's machine was a multiple-needle quilting machine. The threaded needles penetrated the fabric from below and were retracted, leaving the loops on the surface. A thread was drawn through the loops to produce what the inventor termed a chain. The first two stitches were twisted before insertion into the next two, producing a type of twisted lockstitch. The mechanism for feeding the cloth was faulty, however, and the inventor himself stated in the specifications that much remained to perfect and simplify it before its general application. (This machine was illustrated [fig. 10] in the *Sewing Machine Times*, October 25, 1907, and mistakenly referred to as the 1814 model.) Madersperger realized no financial gain from either venture and died in a poorhouse in 1850.

The first efforts of the 1840s reflected the work of the earlier years. In England, Edward Newton and Thomas Archbold invented and patented a machine on May 4, 1841, for tambouring or ornamenting the backs of gloves. Their machine used a hook on the upper surface to catch the loop of thread, but an eye-pointed needle from underneath was used to carry the thread up through the fabric. The machine was designed to use three needles for three rows of chainstitching, if required. Although the machine was capable of stitching two fabrics together, it was never contemplated as a sewing machine in the present use of the term. Their British patent 8,948 stated it was for "improvements in producing ornamental or tambour work in the manufacture of gloves."

The earliest American patent specifically recorded as a sewing machine was U.S. patent 2,466, issued to John J. Greenough on February 21, 1842. His machine was a short-thread model that made both the running stitch and the backstitch. It used the two-pointed needle, with eye at mid-length, which was passed back and forth through the material by means of a pair of pincers on each side of the seam. The pincers opened and closed automatically. The material to be sewn was held in clamps which moved it forward between the pincers to form a running stitch or moved it alternately backward and forward to produce a backstitch. The clamps were attached to a rack that automatically fed the material at a predetermined rate according to the length of stitch required. Since the machine was designed for leather or other hard material, the needle was preceded by an awl, which pierced a hole. The machine had a weight to draw out the thread and a stop-motion to stop the machinery when a thread broke or became too short. The needle was threaded with a short length of thread and required frequent refilling. Only straight seams could be stitched. The feed was continuous to the length of the rack bar; then it had to be reset. The motions were all obtained from the revolution of a crank. It is not believed that any machines, other than the patent model (fig. 11), were ever made. Greenough was with the Patent Office from 1837 to 1841.

In the succeeding year, on March 4, 1843, Benjamin W. Bean received the second American sewing-machine patent, U.S. patent 2,982. Like Greenough's, this machine made a running stitch, but by a different method. In Bean's machine the fabric was fed between the teeth of a series of gears. Held in a

---

[25] The rebuilt machine, according to a letter to the author from B. F. Thompson of the Singer company, is believed to have been one of the machines lost in a Singer factory fire at Elizabethport, N.J., in 1890.

[26] Op. cit. (footnote 24).

Figure 11.—Greenough's patent model, 1842. (Smithsonian photo 45525–G.)

groove in the gears was a peculiarly shaped needle bent in two places to permit it to be held in place by the gears and with a point at one end and the eye at the opposite end, as in a common hand needle. The action of the gears caused the fabric to be forced onto and through the threaded needle. Indefinite straight seams could be stitched as the fabric was continuously forced off the needle by the turning gears (fig. 12). A screw clamp held the machine to a table or other work surface. Machines of this and similar types reportedly had some limited usage in the dyeing and bleaching mills,[27] where lengths of fabric were stitched together before processing. Improved versions of Bean's machine were to be patented in subsequent years in England and America. The same principle was also used in home machines two decades later.

The third sewing-machine patent on record in the United States Patent Office is patent 3,389 issued on December 27, 1843, to George H. Corliss, better remembered as the inventor and manufacturer of the Corliss steam engine. It was his interest in the sewing machine, however, that eventually directed his attention to the steam engine.

Corliss had a general store at Greenwich, New York. A customer's complaint that the boots he had purchased split at the seams made Corliss wonder why someone had not invented a machine to sew stronger seams than hand-sewn ones. He considered the problem of sewing leather, analyzing the steps required to make the saddler's stitch, one popularly used in boots and shoes. He concluded that a sewing machine to do this type of work must first perforate the leather, then draw the threads through the holes, and finally secure the stitches by pulling the threads tight. The machine Corliss invented (fig. 13) was of the same general type as Greenough's, except that two two-pointed needles were required to make the saddler's stitch. This stitch was composed of two running stitches made simultaneously, one from each side.[28] The machine used two awls to pierce the holes through which the needles passed; finger levers

---

[27] Edward H. Knight, *Sewing Machines*, vol. 3 of *Knight's American Mechanical Dictionary*.

[28] A seam using the saddler's stitch appears as a neat line of touching stitches on both sides. Even when made by hand, it is sometimes misidentified by the casual observer as the lockstitch because of the uniformity of both sides. If the saddler's stitch was formed of threads of two different colors, the even stitches on one side of the seam and the odd stitches on the reverse side would be of one color, and vice versa.

approached from opposite sides, seized the needles, pulled the threads firmly, and passed the needles through to repeat the operation. The working model that Corliss completed could unite two pieces of heavy leather at the rate of 20 stitches per minute.

Corliss, lacking capital, went to Providence, Rhode Island, in 1844 to secure backers. After months without success, he was forced to abandon the sewing machine and accept employment as a draftsman and designer. Though he considered himself a failure, this change of employment placed him on the threshold of his more rewarding life work, improvement of the steam engine.[29]

On July 22, 1844, James Rodgers was granted U.S. patent 3,672, the fourth American sewing-machine patent. The patent model is not known to be in existence, but this machine was of minor importance for it offered only a negligible change in the Bean running-stitch machine. The same corrugated gears were used but were placed in different positions so that one bend in the needle was eliminated. When Bean secured a reissue of his patent in 1849, he had adapted it to use a straight needle. Rodgers' machine is not known to have had any commercial success, although this type of machine experienced a brief period of popularity. By the early 1900s, however, the running-stitch machine was so little known that when one was illustrated in the *Sewing Machine Times* in 1907 [30] it excited more curiosity than any of the other early types.

Figure 12.—BEAN'S PATENT MODEL, 1843. (Smithsonian photo 42490–C.)

On December 7, 1844, the same year that Rodgers secured his American patent, John Fisher and James Gibbons were granted British patent 10,424 for "certain improvements in the manufacture of figured or ornamental lace, or net, or other fabrics." From this superficial description of its work, the device might seem to be just another tambouring machine. It was not. Designed specifically for ornamental stitching, the machine made a two-thread stitch using an eye-pointed needle and a shuttle.[31] Several sets of needles and shuttles worked simultaneously. The needles were secured to a needlebar placed beneath the fabric. The shuttles were pointed at both ends to pass through each succeeding new loop formed by the needles. Each shuttle was activated by two vibrating arms worked by cams. Each needle was curved in the form of a bow, and in addition to the eye at the point each also had a second eye at the bottom of the curve. The shape of the needle

---

[29] *The Life and Works of George H. Corliss*, privately printed for Mary Corliss by the American Historical Society, 1930. The Corliss family records were turned over to the Baker Library, Harvard University. In a letter addressed to this author by Robert W. Lovett of the Manuscripts Division on August 2, 1954, it was reported that there was a record on their Corliss card to the effect that a model of his sewing machine, received with the collection, was turned over to the Massachusetts Institute of Technology; however, Mr. Lovett also stated that from a manuscript memoir of Mr. Corliss that it would seem that he developed only the one machine—the patent model. In a letter dated November 15, 1954, Stanley Backer, assistant professor of mechanical engineering, stated that after extensive inquiries they were unable to locate the model at M.I.T. In 1964, Dr. Robert Woodbury, of M.I.T., turned over to the Smithsonian Institution the official copies of the Corliss drawings and the specifications which had been awarded to the inventor by the Patent Office. It is possible that this may have been the material noted on the Harvard University card as having been transferred to M.I.T.

[30] *Sewing Machine Times* (July 10, 1907), vol. 26, no. 858, p. 1.

[31] This is the earliest known patent using the combination of an eye-pointed needle and a shuttle to form a stitch.

15

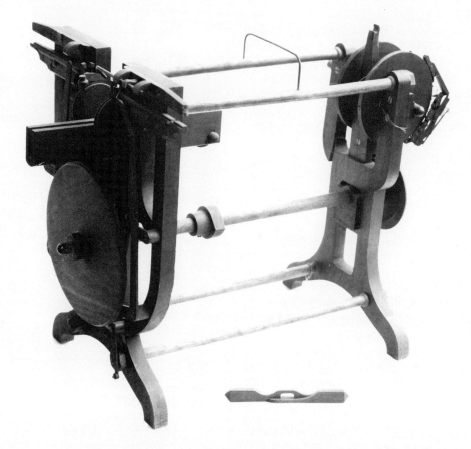

Figure 13.—Corliss' patent model, 1843. The piece of wood in the foreground is an enlarged model of the needle. (Smithsonian photo 42490.)

together with the position of the eyes permitted the pointed shuttle, carrying the second thread, to pass freely through the loop in the ascending needle thread. The fabric was carried by a pair of cloth rollers, capable of sliding in a horizontal plane in both a lateral and a lengthwise direction. These combined movements were sufficient to enable the operator to produce almost every embroidered design. The ornamenting, which might be a yarn, cord, or gimp, was carried by the shuttle thread. There was no tension on the shuttle thread, which was held in place by the thread from the needle. The stitch produced was a form of couching.[32] It was in no sense a lockstitch. Fisher, who was the inventor, readily admitted at a later date that he had not had the slightest idea of producing a sewing machine, in the utilitarian meaning of the term. Although it has not been established that this machine was ever put into practical operation, Fisher's invention was to have a far-reaching effect on the development of the sewing machine in England.

[32] In embroidery, couching is the technique of laying a decorative thread on the surface of the fabric and stitching it into place with a second less-conspicuous thread.

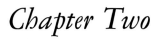

Figure 14.—Howe's prepatent model of 1845, and the box used by the inventor to carry the machine to England in 1847. (Smithsonian photo 45506–B.)

Elements of a Successful Machine

THE REQUIREMENTS for producing a successful, practical sewing machine were a support for the cloth, a needle to carry the thread through the fabric and a combining device to form the stitch, a feeding mechanism to permit one stitch to follow another, tension controls to provide an even delivery of thread, and the related mechanism to insure the precise performance of each operation in its proper sequence. Weisenthal had added a point to the eye-end of the needle, Saint supported the fabric by placing it in a horizontal position with a needle entering vertically, Duncan successfully completed a chainstitch for embroidery purposes, Chapman used a needle with an eye at its point and did not pass it completely through the fabric, Krems stitched circular caps with an eye-pointed needle used with a hook to form a chainstitch, Thimmonier used the hooked needle to form a chainstitch on a fabric laid horizontally, and Hunt created a new stitch that was more readily adapted to sewing by machine than the hand stitches had been, but, although each may have had the germ of an idea, a successful machine had not evolved. There were to be hundreds of patents issued in an attempt to solve these and the numerous minor problems that would ensue. But the problems were solved. And, in spite of its Old World inception, the successful sewing machine can be credited as an American invention.

Although the invention of the practical sewing machine, like most important inventions, was a many-man project, historians generally give full credit to Elias Howe, Jr. Though such credit may be overly generous, Howe's important role in this history cannot be denied.

Elias Howe, Jr., was born on a farm near Spencer, Massachusetts, but he left home at an early age to learn the machinist's trade.[33] After serving an apprenticeship in Lowell, he moved to Boston. In the late 1830s, while employed in the instrument shop of Ari Davis, Howe is reported to have overheard a discussion concerning the need for a machine that would sew. In 1843, when illness kept him from his job for days at a time, he remembered the conversation and the promises of the rich reward that reputedly awaited the successful inventor. Determined to invent such a machine, he finally managed to produce sufficient results to interest George Fisher in buying a one-half interest in his proposed invention. By May 1845, Howe's machine (fig. 14) was used to sew the principle seams of two woolen suits for men's clothing. He continued to demonstrate his machine but found that interest was, at best, indifferent.

Nevertheless, Howe completed a second machine (fig. 15), which he submitted with his application for a patent. The fifth United States patent (No. 4,750) for a sewing machine was issued to him on September 10, 1846. The machine used a grooved and curved eye-pointed needle carried by a vibrating arm, with the needle supplied with thread from a spool. Loops of thread from the needle were locked by a thread carried by a shuttle, which was moved

---

[33] See biographical sketch, pp. 218–221.

through the loop by means of reciprocating drivers. The cloth was suspended in a vertical position, impaled on pins projecting from a baster plate, which moved intermittently under the needle by means of a toothed wheel. The length of each stitching operation depended upon the length of the baster plate, and the seams were necessarily straight. When the end of the baster plate reached the position of the needle, the machine was stopped. The cloth was removed from the baster plate, which was moved back to its original position. The cloth was moved forward on the pins, and the seam continued.

In his patent specifications, Howe claimed the following:

> 1. The forming of the seam by carrying a thread through the cloth by means of a curved needle on the end of a vibrating arm, and the passing of a shuttle furnished with its bobbin, in the manner set forth, between the needle and the thread which it carried, under combination and arrangement of parts substantially the same with that described.
>
> 2. The lifting of the thread that passes through the needle-eye by means of the lifting-rod, for the purpose of forming a loop of loose thread that is to be subsequently drawn in by the passage of the shuttle, as herein fully described, said lifting-rod being furnished with a lifting pin, and governed in its motion by the guide-pieces and other devices, arranged and operating substantially as described.
>
> 3. The holding of the thread that is given out by the shuttle, so as to prevent its unwinding from the shuttle-bobbin after the shuttle has passed through the loop, said thread being held by means of the lever or slipping-piece, as herein made known, or in any other manner that is substantially the same in its operation and result.
>
> 4. The manner of arranging and combining the small lever with the sliding box, in combination with the spring-piece, for the purpose of tightening the stitch as the needle is retracted.
>
> 5. The holding of the cloth to be sewed by the use of a baster-plate furnished with points for that purpose, and with holes enabling it to operate as a rack in the manner set forth, thereby carrying the cloth forward and dispensing altogether with the necessity of basting the parts together.

The five claims, which were allowed Howe in his patent, have been quoted to show that he did not claim the invention of the eye-pointed needle, for which he has so often been credited. The court judgment [34]

Figure 15.—Howe's patent model, 1846. (Smithsonian photo 45525–B.)

that upheld Howe's claim to his patented right to control the use of the eye-pointed needle in combination with a shuttle to form a lockstitch was mistakenly interpreted by some as verifying control of the eye-pointed needle itself.

After patenting his invention, Howe spent three discouraging years in both the United States and in England trying to interest manufacturers in building his sewing machine, under license. Finally, for £250 sterling, he sold the British patent rights to William Thomas and further agreed to adapt the machine to Thomas' manufacture of umbrellas and corsets.[35] This did not prove to be a financial success for Howe and by 1849 he was back in the United States, once again without funds.

On his return, Howe was surprised to find that other inventors were engaged in the sewing-machine

---

[34] *In the Matter of the Application of Elias Howe, Jr. for an Extension of His Sewing Machine Patent Dated September 10, 1846,* New York, 1860, with attachments A and B, U.S. Patent Office. [L.C. call no. TJ 1512.H6265]

[35] It is interesting to note that when William Thomas applied for the British patent of the Howe machine (issued Dec. 1, 1846), the courts would not allow the claim for the combination of the eye-pointed needle and shuttle to form a stitch, due to the Fisher and Gibbons patent of 1844. For more details on Howe's years in England, see his biographical sketch, pp. 218–221.

Figure 16.—An enlargement of the stitching area. (Smithsonian photo 45525–B.)

problem and that sewing machines were being manufactured for sale. The sixth United States sewing-machine patent (No. 5,942) had been issued to John A. Bradshaw on November 28, 1848, for a machine specifically stated as correcting the defects in the E. Howe patent. Bradshaw did not purport that his machine was a new invention. His specifications read:

> The curved needle used in Howe's machine will not by itself form the loop in the thread, which is necessary for the flying bobbin, with its case, to pass through, and has, therefore, to be aided in that operation by a lifting-pin, with the necessary mechanism to operate it. This is a very bungling device, and is a great incumbrance to the action of the machine, being an impediment in the way of introducing the cloth to be sewed, difficult to keep properly adjusted, and very frequently gets entangled between the thread and the needle, by which the latter is frequently broken. This accident happens very often, not withstanding all the precaution which it is possible for the most careful operator to exercise; and inasmuch as the delay occasioned thereby is very considerable, and the needles costly and difficult to replace, it is therefore very important that their breaking in this manner be prevented, which in my machine is done in the most effectual manner by dispensing with the lifting-pin altogether, the loop for the flying bobbin to pass through being made with certainty and of the proper form by means of my angular needle moved in a particular manner just before the flying-bobbin case is thrown. The shuttle and its bobbin for giving off the thread in Howe's machine are very defective . . . my neat and simple bobbin-case . . . gives off its thread with certainty and uniformity . . . . The baster-plate in the Howe machine is very inconvenient and troublesome . . . in my machine . . . the clamp . . . is a very simple and efficient device. . . . The Howe machine is stationary, and the baster-plate or cloth-holder progressive. The Bradshaw machine is progressive and the cloth-holder stationary.

Figure 17.—MOREY AND JOHNSON sewing machine, 1849. Below: The machine is marked with the name of its maker, Safford & Williams. The number 49 is a serial number. Missing parts have been replaced with plastic. (Smithsonian photo 48400; brass plate: 48400–H.)

Bradshaw's patent accurately described some of the defects of the Howe machine, but other inventors were later to offer better solutions to the problems.

Although the Bradshaw machine was not in current manufacture, a machine based on it received the seventh United States sewing-machine patent. Patent 6,099 was issued to Charles Morey and Joseph B. Johnson on February 6, 1849. Their machine (fig. 17) was being offered for sale even before the patent was issued.

This was the first American patent for a chainstitch machine. The stitch was made by an eye-pointed needle carrying the thread through the fabric; the thread was detained by a hook until the loop was enchained by the succeeding one. The fabric was held vertically by a baster plate in a manner similar to the Howe machine. Although not claimed in the patent description, the Morey and Johnson machine also had a bar device for stripping the cloth from the needle. This bar had a slight motion causing a yielding pressure to be exerted on the cloth. Although the patent was not granted until February 6, 1849, the application had been filed in April of the previous year. The machine was featured in the *Scientific American* on January 27, 1849 (fig. 18):

> Morey and Johnson Machine—These machines are very accurately adjusted in all their parts to work in harmony, without this they would be of no use. But they are now used in most of the Print Works and Bleach Works in New England, and especially by the East Boston Flour Company. It sews about one yard per minute, and we consider it superior to the London Sewing Machine the specification of which is in our possession. It [Morey and Johnson] is more simple—and this is a great deal. . . . The price of a machine and right to use $135.[36]

An improvement in the Morey and Johnson machine was patented by Jotham S. Conant for which he was issued a patent on May 8, 1849. Conant's machine offered a slight modification of the cloth bar and of the method of keeping the cloth taut during the stitching operation. No successful use of it is known.

A second improvement of the Morey and Johnson patent was also issued on May 8, 1849; this United States patent (No. 6,439) was to John Bachelder for the first continuous, but intermittent, sewing mechanism. As shown in the patent model (fig. 19), his clothholder consisted of an endless belt supported by and running around three or any other suitable number of cylindrical rollers. A series of pointed wires projected from the surface of the belt near the edge immediately adjacent to the needle. The wires could be placed at regular or irregular distances as required. The shaft of one of the cylindrical

---

[36] The machine referred to as the London Sewing Machine is the British patent of the Thimonnier machine. This patent was applied for by Jean Marie Magnin and was published by *Newton's London Journal*, vol. 39, p. 317, as Magnin's invention.

Figure 18.—A MOREY AND JOHNSON sewing machine as illustrated in *Scientific American*, January 27, 1849. (Smithsonian photo 45771.)

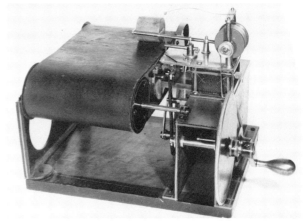

Figure 19.—BACHELDER'S PATENT MODEL, 1849. (Smithsonian photo 45572).

rollers, which supported the endless clothholder, carried a ratchet wheel advanced by the action of a pawl connected to the end of the crankshaft by a small crankpin, whose position or distance from the axis of rotation of the shaft could be adjusted.

By this adjustment the extent of the vertical travel of the impelling pawl was regulated to control the length of the stitch. A spring catch kept the ratchet wheel in place at the end of each forward rotation of the wheel by the pawl. A roller placed over the endless belt at its middle roller pressed the cloth onto the wire points. A curved piece of metal was bent over and down upon the top of the belt so that the cloth, as it was sewed, was carried toward and against the piece by the belt. The cloth rose upon and over the piece and was separated from the points. When the machine was in motion the cloth was carried forward, passed under the needle, was stitched, and finally, passed the separator and off the belt. A vertically reciprocating, straight, eye-pointed needle, a horizontal supporting surface, and a yielding cloth presser were all used, but none were claimed as part of the patent. These were later specifically claimed in reissues of this patent. Bachelder's one specific claim, the endless feed belt, was not limited to belt feeding only. As he explained in the patent, a revolving table or a cylinder might be substituted.

Bachelder did not manufacture machines, but his patent was sold in the mid-1850s to I. M. Singer.[37] It eventually became one of the most important patents to be contributed to the "Sewing-Machine Combination," a patent pool, which is discussed in more detail on pages 41 and 42.

While new ideas and inventors continued to provide the answers to some of the sewing-machine problems, Elias Howe began a series of patent suits to sustain the rights that he felt were his. Since his interest had never been in constructing machines for sale, it was absolutely essential for Howe to protect his royalty rights in order to realize any return from his patent. He was reported[38] to have supervised the construction of 14 sewing machines at a shop[39] on Gold Street in New York toward the close of 1850. Sworn contemporary testimony indicates that the machines were of no practical use.[40] Elias stated, in

---

[37] The exact date is not known; however, it was prior to 1856 as the patent was included in the sewing-machine patent pool formed that year.

[38] JAMES PARTON, *History of the Sewing Machine*, p. 12, (originally published in the *Atlantic Monthly*, May 1867), later reprinted by the Howe Machine Company as a separate.

[39] *Sewing Machine Times* (Feb. 25, 1907), vol. 17, no. 382, p. 1, "His [Bonata's] shop was on Gold Street, New York, near the Bartholf shop, where Howe was building some of his early machines."

[40] *Sewing Machine News*, vol. 3, no. 5, p. 5, Sept. 1881–Jan. 1882. "History of the Sewing Machine."

Figure 20.—BLODGETT & LEROW SEWING MACHINE, 1850, as manufactured by A. Bartholf, New York; the serial number of the machine is 19. The brass plate, serial number 119, at right, is from a later style of the same type of machine. (Smithsonian photo 48440–D; brass plate: 48440–K.)

his application for his patent extension,[41] that he made only one machine in 1850–51. In 1852 he advertised [42] territorial rights and machines, but apparently did not realize any financial success until he sold a half interest in his patent to George Bliss in November 1852.[43] Bliss later began manufacturing machines that he initially sold as "Howe's Patent"; however, these machines were substantially different from the basic Howe machine.

On May 18, 1853, Elias Howe granted his first royalty license to Wheeler, Wilson, Company. Within a few months licenses were also granted to Grover & Baker; A. Bartholf; Nichols & Bliss; J. A. Lerow; Woolridge, Keene, and Moore; and A. B. Howe, the brother of Elias. These licenses granted the manufacturer the right to use any part of the Howe patent,[44] but it did not mean that the machines were Elias Howe machines. When a royalty license was paid, the patent date and sometimes the name was stamped onto the machine. For this reason, these machines are sometimes mistakenly thought to be Elias Howe machines. They are not.

Howe was also prevented from manufacturing a practical machine unless he paid a royalty to other inventors. Three of the major manufacturers and Howe resolved their differences by forming the "Sewing Machine Combination." Although Howe did not enter the manufacturing competition for many years, he profited substantially from the royalty terms of the combination. In 1860, he applied for and received a seven-year extension on his patent.

There were Howe family machines for sale during this period, but these were the ones that Amasa Howe had been manufacturing since 1853. The machine was an excellent one and received the highest medal for sewing machines, together with many flattering testimonials, at the London International Exhibition in 1862. After the publication of this award the demand for (Amasa) Howe sewing

---

[41] Op. cit. (footnote 34).
[42] *New York Daily Tribune*, Jan. 15, 1852, p. 2.
[43] See Howe's biographical sketch, pp. 218–221.
[44] Op. cit. (footnote 34). Attachments A and B are copies of Judge Sprague's decisions.

Figure 21.—BLODGETT & LEROW SEWING MACHINE, 1850, stamped with the legend "Goddard, Rice & Co., Makers, Worcester, Mass." and the serial number 37. Below: An original brass plate marked "No. 38"; this plate fits the machine perfectly. (Smithsonian photo 48440-E; brass plate: 48440-J.)

machines was greatly increased at home and abroad. Elias took this opportunity to gain entry into the manufacturing business by persuading Amasa to let him build a factory at Bridgeport, Connecticut, and manufacture the (Amasa) Howe machines. Two years passed before the factory was completed, and Amasa's agents were discouraged. The loss could have been regained, but the machines produced at Bridgeport were not of the quality of the earlier machines. Amasa attempted to rebuild the Bridgeport machines, but finally abandoned them and resumed manufacturing machines in New York under his own immediate supervision.[45] Elias formed his own company and continued to manufacture sewing machines. In 1867 he requested a second extension of his patent, but the request was refused. Elias Howe died in October of the same year.

Meanwhile, another important sewing machine of a different principle had also been patented in 1849. This was the machine of Sherburne C. Blodgett, a tailor by trade, who was supported financially by John A. Lerow. United States patent 6,766 was issued to both men on October 2, 1849. In the patent, the machine was termed as "our new 'Rotary Sewing Machine'." The shuttle movement was continuous, revolving in a circle, rather than reciprocating as in the earlier machines. Automatic tension was initiated, restraining the slack thread from interference with the point of the needle.

The Blodgett and Lerow machine was built by several shops. One of the earliest was the shop of Orson C. Phelps on Harvard Place in Boston. Phelps took the Blodgett and Lerow machine to the sixth exhibition of the Massachusetts Charitable Mechanics Association in September 1850 and won a silver medal and this praise, "This machine performed admirably; it is an exceedingly ingenious and compact machine, able to perform tailor's sewing beautifully and thoroughly."[46] Although Phelps had

---

[45] *Sewing Machine Journal* (July 1887), pp. 93–94.

[46] *Report of the Sixth Exhibition of the Massachusetts Charitable Mechanics Association, in the City of Boston, September 1850* (Boston, 1850).

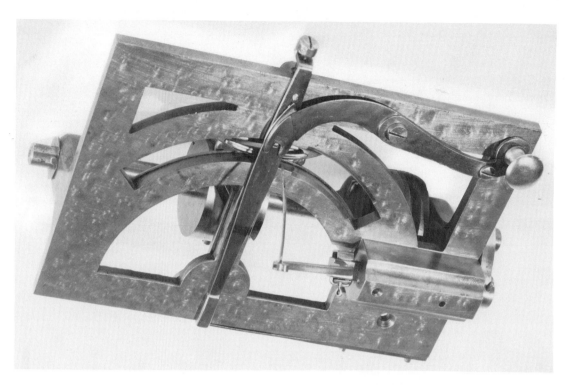

Figure 22.—WILSON'S PREPATENT MODEL for his reciprocating-shuttle machine, 1850. (Smithsonian photo 45525–A.)

won the earliest known premium for a sewing machine, and although the machine was produced commercially to a considerable extent (figs. 20 and 21), one outstanding flaw in its operation could not be overlooked. As the shuttle passed around the six-inch circular shuttle race, it put a twist in the thread (or took one out if the direction was reversed) at each revolution. This caused a constant breaking of the thread, a condition that could not be rectified without changing the principle of operation. Such required changes were later to lead I. M. Singer, another well-known name, into the work of improving this machine.

Also exhibited at the same 1850 mechanics fair was the machine of Allen B. Wilson. Wilson's machine received only a bronze medal, but his inventive genius was to have a far greater effect on the development of the practical sewing machine than the work of Blodgett and Lerow. A. B. Wilson [47] was one of the ablest of the early inventors in the field of mechanical stitching, and probably the most original.

Wilson, a native of Willett, New York, was a young cabinetmaker at Adrian, Michigan, in 1847 when he first conceived of a machine that would sew. He was apparently unaware of parallel efforts by inventors in distant New England. After an illness, he moved to Pittsfield, Massachusetts, and pursued his idea in earnest. By November 1848 he had produced the basic drawings for a machine that would make a lockstitch. The needle, piercing the cloth, left a loop of thread below the seam. A shuttle carrying a second thread passed through the loop, and as the tension was adjusted a completed lockstitch was formed (fig. 22). Wilson's shuttle was pointed on both ends to form a stitch on both its forward and backward motion, a decided improvement over the shuttles of Hunt and Howe, which formed stitches in only one direction. After each stitch the cloth was advanced for the next stitch by a sliding bar against which the cloth was held by a stationary presser. While the needle was still in the cloth and holding it, the sliding bar returned for a fresh grip on the cloth.

Wilson made a second machine, on the same principle, and applied for a patent. He was approached by the owners of the Bradshaw 1848 patent, who

---

[47] See biographical sketch, pp. 221–222.

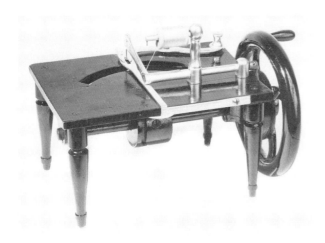

Figure 23.—WILSON'S PATENT MODEL, 1850. (Smithsonian photo 45504–H.)

Figure 24.—WILSON'S PREPATENT MODEL for his rotary hook, 1851. (Smithsonian photo 45506–E.)

claimed control of the double-pointed shuttle. Although this claim was without justification, as can be seen by examining the Bradshaw patent specifications, Wilson did not have sufficient funds to fight the claim. In order to avoid a suit, he relinquished to A. P. Kline and Edward Lee, a one-half interest in his U.S. patent 7,776 which was issued on November 12, 1850 (fig. 23).

Inventor Wilson had been associated with Kline and Lee (E. Lee & Co.) for only a few months, when, on November 25, 1850, he agreed to sell his remaining interest to his partners for $2,000. He retained only limited rights for New Jersey and for Massachusetts. The sale was fruitless for the inventor, as no payment was ever made. How much money E. E. Lee & Co. realized from the Wilson machine is difficult to determine, but they ran numerous ads in the 1851 and 1852 issues of *Scientific American*. A typical one reads:

> A. B. Wilson's Sewing Machine, justly allowed to be the cheapest and best now in use, patented November 12, 1850; can be seen on exhibition at 195 and 197 Broadway (formerly the Franklin House, Room 23, third floor) or to E. E. Lee & Co., Earle's Hotel. Rights for territory or machines can be had by applying to George R. Chittenden, Agent.[48]

Figure 25.—WILSON'S ROTARY-HOOK PATENT MODEL, 1851. (Smithsonian photo 45505–B.)

Another reads:

> A. B. Wilson's Sewing Machine . . . the best and only practical sewing machine—not larger than a lady's work box—for the trifling sum of $35.[49]

Wilson severed relations with Lee and Kline in early 1851 shortly after meeting Nathaniel Wheeler, who was to become his partner in a happier, more

---

[48] *Scientific American* (Dec. 6, 1851), vol. 7, no. 12, p. 95.

[49] Ibid. (Sept. 20, 1851), vol. 7, no. 1, p. 7.

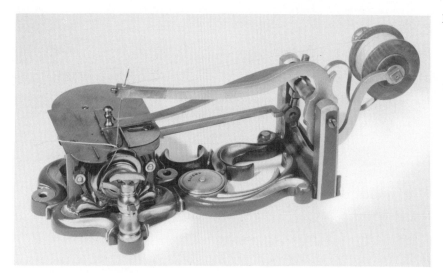

Figure 26.—WILSON's stationary-bobbin patent model, 1852; a commercial machine was used since Wheeler, Wilson, Co. had begun manufacturing machines the previous year. (Smithsonian photo 45504–B.)

profitable enterprise involving the sewing machine.

Wilson, with his two partners, was occupying a room in the old Sun Building at 128 Fulton Street, when Wheeler, on a business trip to New York City, learned of the Wilson sewing machine. Wheeler examined the machine, saw its possibilities, and at once contracted with E. Lee & Co. to make 500 of them. At the same time he engaged Wilson to go with him to Watertown, Connecticut, to perfect the machine and supervise its manufacture. Meanwhile, Wilson had been working on a substitute for the shuttle. He showed his model of the device, which became known as the rotary hook, to Wheeler who was so convinced of its superiority that he decided to develop this new machine and leave Wilson's first machine to the others, who, by degrees, had become its owners.

Wilson now applied all his effort to improving the rotary hook, for which he received his second patent on August 12, 1851 (figs. 24 and 25). Wheeler, his two partners Warren and Woodruff, and Wilson now formed a new copartnership—Wheeler, Wilson, Company. They began the manufacture of the machines under the patent, which combined the rotary hook and a reciprocating bobbin. The rotary hook extended or opened more widely the loop of the needle thread, while a reciprocating bobbin carried its thread through the extended loop. To avoid litigation which the reciprocating bobbin might have caused, Wilson contrived his third outstanding invention—the stationary bobbin. This was a feature of the first machine produced by the new company in 1851, though the patent for the stationary bobbin was not issued until June 15, 1852 (fig. 26).

In all reciprocating-shuttle machines a certain loss of power is incurred in driving forward, stopping, and bringing back the shuttle at each stitch; also, the machines are rather noisy, owing to the striking of the driver against the shuttle at each stroke. These objections were removed by Wilson's rotary hook and stationary bobbin. The locking of the needle thread with the bobbin thread was accomplished, not by driving a shuttle through the loop of the needle thread, but by passing that loop under the bobbin. The driving shaft carried the circular rotary hook, one of the sewing machine's most beautiful contrivances. The success of the machine is indicated in an article that appeared in the June 1853 issue of *Scientific American:*

> There are 300 of these machines now in operation in various parts of the country, and the work which they can perform cannot be surpassed.... The time must soon come when every private family that has much sewing to do, will have one of these neat and perfect machines; indeed many private families have them now.... The price of one all complete is $125; every machine is made under the eye of the inventor at the company's machine shop, Watertown, Connecticut, so that every one is warranted ... agreement between Mr. Howe and Messrs. Wheeler, Wilson & Co., so every customer will be perfectly protected ....[50]

---

[50] Ibid. (June 4, 1853), vol. 7, no. 38, p. 298

Figure 27.—WILSON's four-motion-feed patent model, 1854, is not known to be in existence; this is a commercial machine of the period. The plate is stamped "A. B. Wilson, Patented Aug. 12, 1851, Watertown, Conn., No. 1. . . ." (Smithsonian photo 45504.)

This agreement was important to sales, as Elias Howe was known to have sued purchasers of machines, as well as rival inventors and companies.

The business was on a substantial basis by October 1853, and a stock company was formed under the name of Wheeler & Wilson Manufacturing Company.[51] A little more than a year later, on December 19, 1854, Wilson's fourth important patent (U.S. patent 12,116)—for the four-motion cloth feed—was issued to him (fig. 27). In this development, the flat-toothed surface in contact with the cloth moved forward carrying the cloth with it; then it dropped a little, so as not to touch the cloth; next it moved backward; then in the fourth motion it pushed up against the cloth and was ready to repeat the forward movements. This simple and effective feed method is still used today, with only minor modifications, in almost every sewing machine. This feed with the rotary hook and the stationary circular-disk bobbin, completed the essential features of Wilson's machine. It was original and fundamentally different from all other machines of that time.

The resulting Wheeler and Wilson machine made a lockstitch by means of a curved eye-pointed needle carried by a vibrating arm projecting from a rock shaft connected by link and eccentric strap with an eccentric on the rotating hook shaft. This shaft had at its outer end the rotary hook, provided with a point adapted to enter the loop of needle thread. As the hook rotated, it passed into and drew down the loop of needle thread, which was held by means of a loop check, while the point of the hook entered a new loop. When the first loop was cast off—the face of the hook being beveled for that purpose—it was drawn upward by the action of the hook upon the loop through which it was then passing. During the rotation of the hook each loop was passed around a disk bobbin provided with the second thread and serving the part of the shuttle in other machines. The four-motion feed was actuated in this machine by means of a spring bar and a cam in conjunction with the mandrel.

From the beginning, Wheeler and Wilson had looked beyond the use of the sewing machine solely by manufacturers and had seen the demand for a light-running, lightweight machine for sewing in the home. Wilson's inventions lent themselves to this design, and Wheeler and Wilson led the way to the introduction of the machine as a home appliance. Other manufacturers followed.

When the stock company was formed, Mr. Wilson retired from active participation in the business at his own request. His health had not been good, and a nervous condition made it advisable for him to be freed from the responsibility of daily routine. During this period Wilson's inventive contributions to the sewing machine continued as noted, and in addition he worked on inventions concerning cotton picking and illuminating gases.

Wheeler and Wilson's foremost competitor in the

---

[51] J. D. VAN SLYCK, *New England Manufactures and Manufactories*, vol. 2, pp. 672–682.

early years of sewing-machine manufacture was the Singer Company, which overtook them by 1870 and finally absorbed the entire Wheeler & Wilson Manufacturing Company in 1905.

The founder of this most successful 19th-century company was Isaac Singer, a native of Pittstown, New York.[52] Successively a mechanic, an actor, and an inventor, Singer came to Boston in 1850 to promote his invention of a machine for carving printers' wooden type. He exhibited the carving machine in Orson Phelps' shop, where the Blodgett and Lerow machines were being manufactured.

Because the carving machine evoked but little interest, Singer turned his attention to the sewing machine as a device offering considerable opportunity for both improvement and financial reward. Phelps liked Singer's ideas and joined with George Zieber, the publisher who had been backing the carving-machine venture, to support Singer in the work of improving the sewing machine. His improvements in the Blodgett and Lerow machine included a table to hold the cloth horizontally rather than vertically (this had been used by Bachelder and Wilson also), a yielding vertical presser foot to hold the cloth down as the needle was drawn up, and a vertically reciprocating straight needle driven by a rotary, overhanging shaft.

The story of the invention and first trial of the machine was told by Singer in the course of a patent suit sometime later:

> I explained to them how the work was to be fed over the table and under the presser-foot, by a wheel, having short pins on its periphery, projecting through a slot in the table, so that the work would be automatically caught, fed and freed from the pins, in place of attaching and detaching the work to and from the baster plate by hand, as was necessary in the Blodgett machine.
>
> Phelps and Zieber were satisfied that it would work. I had no money. Zieber offered forty dollars to build a model machine. Phelps offered his best endeavors to carry out my plan and make the model in his shop; if successful we were to share equally. I worked at it day and night, sleeping but three or four hours a day out of the twenty-four, and eating generally but once a day, as I knew I must make it for the forty dollars or not get it at all.
>
> The machine was completed in eleven days. About nine o'clock in the evening we got the parts together and tried it; it did not sew; the workmen exhausted with almost unremitting work, pronounced it a failure and left me one by one.
>
> Zieber held the lamp, and I continued to try the machine, but anxiety and incessant work had made me nervous and I could not get tight stitches. Sick at heart, about midnight, we started for our hotel. On the way we sat down on a pile of boards, and Zieber mentioned that the loose loops of thread were on the upper side of the cloth. It flashed upon me that we had forgot to adjust the tension on the needle thread. We went back, adjusted the tension, tried the machine, sewed five stitches perfectly and the thread snapped, but that was enough. At three o'clock the next day the machine was finished. I took it to New York and employed Mr. Charles M. Keller to patent it. It was used as a model in the application for the patent.[53]

The first machine was completed about the last of September 1850. The partners considered naming the machine the "Jenny Lind," after the Swedish soprano who was then the toast of America. It was reported [54] to have been advertised under that name when the machine was first placed on the market, but the name was soon changed to "Singer's Perpendicular Action Sewing Machine" or simply the "Singer Sewing Machine"—a name correctly anticipated to achieve a popularity of its own.

According to the contract made by the partners, the hurriedly built first machine was to be sent to the Patent Office with an application in the name of Singer and Phelps. An application was made between the end of September 1850 and March 14, 1851, as Singer refers to it briefly in the application formally filed on April 16, 1851, stating, "My present invention is of improvements on a machine heretofore invented by me and for which an application is now pending." [55]

---

[52] See biographical sketch, pp. 222–223.

[53] Chester McNeil, *A History of the Sewing Machine* in Union Sales Bulletin, vol. 3, Union Special Sewing Machine Co., Chicago, Illinois, pp. 83–85. 1903.

[54] *Sewing Machine Times* (Aug. 25, 1908), vol. 18, no. 418.

[55] Singer gives this limited description of the first machine, with detailed improvements for which he was then applying for a patent: "In my previous machine, to which reference has been made, the bobbin was carried by the needle-carrier, and hence the motion of the needle had to be equal to the length of thread required to form the loop, which was objectionable, as in many instances this range of motion was unnecessarily long for all other purposes . . . ." Quoted from U.S. patent 8,294 issued to Isaac M. Singer, Aug. 12, 1851. It should be noted that in some instances there was a considerable lapse of time from the date a patent application was made until the patent was issued. In this case the handwritten specifications were dated March 14, 1851, and the formal Patent Office receipt was dated April 16, 1851.

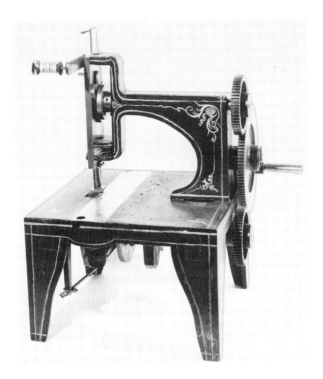

Figure 28.—SINGER'S PATENT MODEL, 1851; a commercial machine was used, bearing the serial number 22. (Smithsonian photo 45572–D.)

Figure 29.—SINGER'S PERPENDICULAR ACTION sewing machine, an engraving from *Illustrated News*, June 25, 1853, which states: "The sewing machine has, within the last two years acquired a wide celebrity, and established its character as one of the most efficient labor saving instruments ever introduced to public notice . . . . We must not forget to call attention to the fact that this instrument is peculiarly calculated for female operatives. They should never allow its use to be monopolized by men." (Smithsonian photo 48091–D.)

In late December 1850 Singer had bought Phelps' interest in the company. Whether the first application was later abandoned by Singer or whether it was rejected is not known,[56] but a patent on the first application was never issued. The final disposition of this first machine has remained a mystery.[57]

A few machines were manufactured in late 1850 and early 1851, and these attracted considerable attention; orders began to be received in advance of production. The pending patent application did not delay the manufacture, and a number of machines were sold before August 12, 1851, when the patent was granted. The patent model is shown in figure 28.[58] It made a lockstitch by means of a straight eye-pointed needle and a reciprocating shuttle. The patent claims, as quoted from the specifications, were as follows:

1. Giving to the shuttle an additional forward motion after it has been stopped to close the loop, as described, for the purpose of drawing the stitch tight, when such additional motion is given at and in combination with

---

[56] If a patent was not approved, for any reason, the records were placed in an "Abandoned File." In 1930 Congress authorized the disposal of the old "Abandoned Files," requiring them to be kept for twenty years only. There are no Singer Company records giving an account of the first patent application.

[57] Its whereabouts was unknown as early as 1908, as stated in the *Sewing Machine Times* (Aug. 25, 1908), vol. 18, no. 418. Models of abandoned patents frequently remained at the Patent Office. Approximately 76,000 models were ruined in a Patent Office fire in 1877. In 1908 over 3000 models of abandoned patents were sold at auction. Either incident could account for the machine's disappearance.

[58] The patent model of 8,294 is a machine that bears the serial number 22; it was manufactured before April 18, 1851, the date it was recorded as received by the Patent Office.

Figure 30.—I. M. Singer & Co. New York showroom of the mid-1850s, as illustrated in *Frank Leslie's Illustrated Newspaper*, August 29, 1857; only manufacturing machines are shown in this illustration. (Smithsonian photo 48091–B.)

the feed motion of the cloth in the reverse direction, and the final upward motion of the needle, as described, so that the two threads shall be drawn tight at the same time, as described.

2. Controlling the thread during the downward motion of the needle by the combination of a friction-pad to prevent the slack above the cloth, with the eye on the needle-carrier for drawing back the thread, for the purposes and in the manner substantially as described.

3. Placing the bobbin from which the needle is supplied with thread on an adjustable arm attached to the frame, substantially as described, when this is combined with the carrying of the said thread through an eye or guide attached to and moving with the needle-carrier, as described, whereby any desired length of thread can be given for the formation of the loop without varying the range of motion of the needle, as described.

The feeding described in the Singer patent was "by the friction surface of a wheel, whose periphery is formed with very fine grooves, the edges of which are slightly serrated, against which the cloth is pressed by a spring plate or pad." Although claimed by the inventor in the handwritten specifications, it was not allowed as original.

The machines manufactured by the Singer company (fig. 29) were duplicates of the patent model. These machines were quite heavy and intended for

Figure 31.—Hunt and Webster's sewing-machine manufactory exhibition and salesroom in Boston, as illustrated in *Ballou's Pictorial*, July 5, 1856; only manufacturing machines are shown. (Smithsonian photo 45771–A.)

manufacturing rather than for family use in the home.

Singer enjoyed demonstrating the machine and showed it to church and social groups and even at circuses; this personal association then encouraged him to improve its reliability and convenience. He developed a wooden packing case which doubled as a stand for the machine and a treadle to allow it to be operated by foot. Because of the dimensions of the packing case, Singer put the pivot of the treadle toward its center, about where the instep of the foot would rest. This produced the heel-and-toe action treadle, a familiar part of the sewing machine until its replacement by the electric motor. Both hands were freed to guide and arrange the cloth that was being stitched. Singer also added a flywheel to smooth out the treadle action and later an iron stand with a treadle wide enough for both feet. The treadle had been in use for two years before a rival pointed out that it might have been patented. To Singer's chagrin it was then too late for patent laws did not permit patenting a device that had been in public use.

A new obstacle appeared in the Singer company's path when Howe demanded $25,000 for infringement of his patent. Singer and Zieber decided to fight, enlisting the legal aid of Edward Clark, a lawyer and financier. Howe's action was opposed on the basis of Hunt's machine of 1834, which they stated had anticipated Howe's invention.

While they were resisting, Howe sued three firms that were using and selling Singer machines. The court order required the selling firms and the purchasers to provide an account of the profits accrued from the sale and the use of the sewing machines and restrained the firms from selling the machines during the pendency of the suit.[59] As a result of this action, a

---

[59] William R. Bagnall, in "Contributions to American Economic History," vol. 1 (1908), MS, Harvard School of Business Library.

Figure 32.—Singer's new Family Sewing Machine, illustration from a brochure dating about 1858 or 1859 which states: "A few months since, we came to the conclusion that the public taste demanded a sewing machine for family purposes more exclusively; a machine of smaller size, and of a lighter and more elegant form; a machine decorated in the best style of art, so as to make a beautiful ornament in the parlor or boudoir; a machine very easily operated, and rapid in working . . . . To supply this public want, we have just produced, and are now prepared to receive orders for, 'Singer's new Family Sewing Machine.'" (Smithsonian photo 48091-H.)

number of Singer's rivals purchased licenses from Howe and advertised that anyone could sell their machines without fear of a suit. This gave them a great competitive advantage, and Singer and Clark [60] decided it was best to seek a settlement with Howe. On July 1, 1854, they paid him $15,000 and took out a license.

In spite of this defeat, the Singer company could

---

[60] Singer purchased Phelps' interest in the company in 1851 and sold it to Edward Clark.

Figure 33.—Singer Family Machine, 1858, head only. (Smithsonian photo 45524-F.)

claim several important improvements to the sewing machine and the acquisition of the patents rights to the Morey and Johnson machine of 1849, which gave them control of the spring or curved arm to hold the cloth by a yielding pressure. Although this point had not been claimed in the 1849 patent, the established principle of patent law allowed that a novel device introduced and used in a patented machine could be covered by a reissue at any time during the life of the patent. Upon becoming owners of the Morey and Johnson patent, Singer applied for a reissue which covered this type of yielding pressure. It was granted on June 27, 1854. The Singer company's acquisition of the Bachelder patent had given them control of the yielding pressure bar also.

Singer's aggressive selling had begun to overcome the public's suspicion of sewing machines. He pioneered in the use of lavishly decorated sewing-machine showrooms when the company offices were expanded in the mid-1850s (fig. 30). These were rich with carved walnut furniture, gilded ornaments, and carpeted floors, places in which Victorian women were not ashamed to be seen. The machines were demonstrated by pretty young women. The total effect was a new concept of selling, and Singer became the drum major of a new and coming industry that had many followers (see fig. 31).

The first, light, family sewing machine by the Singer

Figure 34.—GROVER AND BAKER'S PATENT MODEL, 1851. (Smithsonian photo 32003–G.)

company was not manufactured until 1858 (figs. 32 and 33). Comparatively few of these machines were made as they proved to be too small and light. The men in the shop dubbed the machine "The Grasshopper," but it was officially called the new Family Sewing Machine or the Family Machine.[61] Because of its shape, Singer company brochures of the 1920s referred to it as the Turtleback Machine.

Since the cost of sewing machines was quite high and the average family income was low, Clark suggested the adoption of the hire-purchase plan. Into the American economy thus came the now-familiar installment buying.

Singer and Clark continued to be partners until 1863 when a corporation was formed. At this time Singer decided to withdraw from active work. He received 40 percent of the stock and retired to Paris and later to England, where he died in 1875.

By the mid-1850s the basic elements of a successful, practical sewing machine were at hand, but the continuing court litigation over rival patent rights seemed destined to ruin the economics of the new industry. It was then that the lawyer of the Grover and Baker company, another sewing-machine manufacturer of the early 1850s, supplied the solution. Grover and Baker were manufacturing a machine that was mechanically good, for this early period. William O. Grover was another Boston tailor, who, unlike many others, was convinced that the sewing machine was going to revolutionize his chosen trade. Although the sewing machines that he had seen were not very practical, he began in 1849 to experiment with an idea based on a new kind of stitch. His design was for a machine that would take both its threads from

---

[61] This first, family sewing machine should not be confused in name with a model brought out in the sixties. The name of this first, family machine was in the sense of a new "family" sewing machine. In 1859 a "Letter A" family machine was introduced. Thus in 1865 when the Singer Company brought out another family machine they called it the "New" Family Sewing Machine. Both the first-style Family machine and the Letter A machine are illustrated in *Eighty Years of Progress of the United States* (New York, 1861), vol. 2, p. 417, and discussed in an article, "The Place and Its Tenants," in the *Sewing Machine Times* (Dec. 25, 1908), vol. 27, no. 893.

35

Figure 35.—This Grover and Baker cabinet-style sewing machine of 1856 bears the serial number 5675 and the patent dates February 11, 1851, June 22, 1852, February 22, 1853, and May 27, 1856. (Smithsonian photo 45572–F.)

spools and eliminate the need to wind one thread upon a bobbin. After much experimenting, he proved that it was possible to make a seam by interlocking two threads in a succession of slipknots, but he found that building a machine to do this was a much more difficult task. It is quite surprising that while he was working on this idea, he did not stumble upon a good method to produce the single-thread (as opposed to Grover and Baker's two-thread) chainstitch, later worked out by another. Grover was working so intently on the use of two threads that apparently no thought of forming a stitch with one thread had a chance to develop.

At this time Grover became a partner with another Boston tailor, William E. Baker, and on February 11, 1851, they were issued U.S. patent No. 7,931 for a machine that did exactly what Grover had set out to do; it made a double chainstitch with two threads both carried on ordinary thread spools. The machine (figs. 34 and 35) used a vertical eye-pointed needle for the top thread and a horizontal needle for the underthread. The cloth was placed on the horizontal platform or table, which had a hole for the entry of the vertical needle. When this needle passed through the cloth, it formed a loop on the underside. The horizontal needle passed through this loop forming another loop beyond, which was retained until the redescending vertical needle

Figure 36.—Grover's patent model for the first portable case, 1856. The machine in the case is a commercial machine of 1854, bearing the serial number 3012 and the patent dates "Feby 11, 1851, June 22, 1852, Feby 22, 1853." Powered by a single, foot-shaped treadle that was connected by a removable wooden pitman, it also could be turned by hand. (Smithsonian photo 45525–D.)

enchained it, and the process repeated. The slack in the needle thread was controlled by means of a spring guide. The cloth was fed by feeding rolls and a band.

A company was organized under the name of Grover and Baker Sewing Machine Company, and soon the partners took Jacob Weatherill, mechanic, and Orlando B. Potter, lawyer (who became the president), into the firm. Potter contributed his ability as a lawyer in lieu of a financial investment and handled the several succeeding patents of Grover and Baker. These patents were primarily for mechanical improvements such as U.S. patent No. 9,053 issued to Grover and Baker on June 22, 1852, for devising a curved upper needle and an under looper [62] to form the double-looped stitch which became known as the Grover and Baker stitch.

---

[62] A looper on the underside in place of the horizontal needle.

37

One of the more interesting of the patents, however, was for the box or sewing case for which Grover was issued U.S. patent No. 14,956 on May 27, 1856. The inventor stated "that when open the box shall constitute the bed for the machine to be operated upon, and hanging the machine thereto to facilitate oiling, cleansing, and repairs without removing it from the box." It was the first portable sewing machine (fig. 36).

Though the Grover and Baker company manufactured machines using a shuttle and producing the more common lockstitch, both under royalty in their own name and also for other smaller companies, Potter was convinced that the Grover and Baker stitch was the one that eventually would be used in both family and commercial machines. He, as president, directed the efforts of the company to that end. When the basic patents held by the "Sewing-Machine Combination" (discussed on pp. 41–42) began to run out in the mid-1870s, dissolving its purpose and lowering the selling price of sewing machines, the Grover and Baker company began a systematic curtailing of expenses and closing of branch offices. All the patents held by the company and the business itself were sold to another company.[63] But the members of the Grover and Baker company fared well financially by the strategic move.

The Grover and Baker machine and its unique stitch did not have a great influence on the overall development of the mechanics of machine sewing. The merits of a double-looped stitch—its elasticity and the taking of both threads from commercial spools—were outweighed by the bulkiness of the seam and its consumption of three times as much thread as the lockstitch required. Machines making a similar type of stitch have continued in limited use in the manufacture of knit goods and other products requiring an elastic seam. But, more importantly, Grover and Baker's astute Orlando B. Potter placed their names in the annals of sewing-machine history by his work in forming the "Combination," believed to be the first "trust" of any prominence.

---

[63] Domestic Sewing Machine Company. See *Union Special Sewing Machine Co. Sales Bulletin*, vol. 3, ch. 15, pp. 58–59.

Chapter Three

A PARTIAL STATEMENT FROM RECORDS OF "THE SEWING-MACHINE COMBINATION," SHOWING NUMBER OF SEWING-MACHINES LICENSED ANNUALLY UNDER THE *ELIAS HOWE* PATENT.

Name of Manufacturer.	1853.	1854.	1855.	1856.	1857.	1858.	1859.	1860.	1861.	1862.	1863.	1864.	1865.	1866.
Wheeler & Wilson Mfg. Co...	799	756	1,171	2,210	4,591	7,978	21,306	25,102	18,556	28,202	29,778	40,062	39,157	50,132
I. M. Singer & Co.	810	879	883	2,564	3,630	3,594	10,953	13,000 (a)	16,000 (a)	18,396	....	....	....	....
The Singer Manufacturing Co.	...	....	....	....	....	....	....	....	....	....	20,030	23,632	26,340	30,960
Grover & Baker S. M. Co. ..	657	2,034	1,144	1,952	3,680	5,070	10,280	(b)	(b)	(b)	(b)	(b)	(b)	(b)
A. B. Howe " " ..	...	60	53	47	133	179	921	(b)	(b)	(b)	(b)	(b)	(b)	(b)
Leavitt " " ..	28	217	152	235	195	75	213	(b)	(b)	(b)	(b)	(b)	(b)	(b)
Ladd & Webster " " ..	100	268	73	180	453	490	1,788	(b)	(b)	(b)	(b)	(b)	(b)	(b)
Bartholf " " ..	135	55	31	35	31	203	747	(b)	(b)	(b)	(b)	(b)	(b)	(b)

A PARTIAL STATEMENT SHOWING NUMBER OF SEWING-MACHINES LICENSED ANNUALLY FROM 1867 TO 1876 INCLUSIVE.

Name of Manufacturer.	1867.	1868.	1869.	1870.	1871.	1872.	1873.	1874.	1875.	1876.
The Singer Manufacturing Company...	43,053	59,629	86,781	127,833	181,260	219,758	232,444	241,679	249,852	262,316
Wheeler & Wilson Mfg. Company......	38,055	(b)	78,866	83,208	128,526	174,088	119,190	92,827	103,740	108,997
Grover & Baker Sewing-Machine Co...	32,999	35,000 (a)	35,188	57,402	50,838	52,010	36,179	20,000 (a)	15,000 (a)	....
Weed Sewing-Machine Co....	3,638	12,000	19,687	35,002	39,655	42,444	21,769	20,495	21,993	14,425
Howe Sewing-Machine Co....	11,053	35,000 (a)	45,000 (a)	75,156	134,010	145,000 (a)	90,000 (a)	35,000 (a)	25,000 (a)	109,294
A. B. Howe " "	......	......	......	......	20,051	......	......	......	......	......
B. P. Howe " "	......	......	......	......	......	......	13,919	......	......	......
Willcox & Gibbs Sewing-Machine Co...	14,152	15,000	17,201	28,890	30,127	33,639	15,881	13,710	14,522	12,758
Wilson (W. G.) " "	......	......	......	......	21,153	22,666	21,247	17,525	9,508	......
American B. H. & S. M. Co............	......	......	7,792	14,573	20,121	18,930	14,182	13,529	14,406	17,937
Florence S. M. Co..................	10,534	12,000	13,661	17,660	15,947	15,793	8,960	5,517	4,892	2,978
Shaw & Clark Sewing-Machine Co.....	2,692	3,000	......	......	......	......	......	......	......	......
Gold Medal " " "	......	......	......	8,912	13,562	18,897	16,431	15,214	14,262	7,185
Davis " " "	......	......	......	......	11,568	11,376	8,861	......	......	......
Domestic " " "	......	......	......	10,397	49,554	40,114	22,700	21,452	23,587	
Finkle & Lyon Mfg. Co. and Victor	2,488	2,000	1,339	2,420	7,639	11,901	7,446	6,292	6,103	5,750
Ætna Sewing-Machine Co..........	2,958	3,500	4,548	5,806	4,720	4,262	3,081	1,866	1,447	707
Blees " " "	......	......	......	......	4,557	6,053	3,458	......	......	......
Elliptic " " "	......	......	......	......	4,555	......	......	......	......	......
Empire " " "	3,185	......	......	......	......	......	......	......	......	......
Remington Sewing-Machine Co.......	2,121	5,000	8,700	......	......	......	......	......	......	......
Parham " "	......	......	......	3,560	2,965	4,982	9,183	17,608	25,110	12,716
Bartram & Fanton Mfg. Co........	......	......	1,141	1,766	2,056	......	......	......	......	......
Bartlett Sewing-Machine Co.........	2,958	......	......	......	1,004	1,000	1,000	250	......	......
J. G. Folsom...................	......	......	......	......	614	1,000	......	......	......	......
McKay Sewing-Machine Asso........	......	......	......	......	280	......	......	......	......	......
C. F. Thompson................	......	......	......	129	218	......	......	128	161	102
Union Buttonhole Machine Co........	......	......	......	......	147	......	......	......	......	......
Leavitt Sewing-Machine Co.........	......	......	......	......	124	......	......	......	......	......
Goodspeed & Wyman S. M. Co.......	1,051	1,000	771	......	......	......	......	......	......	......
Keystone Sewing-Machine Co........	2,126	......	......	......	......	2,665	217	37	......	......
Secor " " "	......	......	......	......	......	311	3,430	4,541	1,307	......
Centennial " " "	......	......	......	......	......	514	......	......	......	......

(*a*) Number estimated.   (*b*) No data.

Figure 37.—TABLE OF SEWING-MACHINE STATISTICS. From Frederick G. Bourne, "American Sewing Machines" in *One Hundred Years of American Commerce*, vol. 2, ed. Chauncey Mitchell Depew (New York: D.O. Haines, 1895), p. 530. (Smithsonian photo 42542–A.)

The "Sewing-Machine Combination"

WITH THE BASIC ELEMENTS of a successful sewing machine assembled, the various manufacturers should have been able to produce good machines unencumbered. The court order, however, which restrained several firms from selling Singer machines while the Howe suit was pending, started a landslide; soon Wheeler, Wilson and company, Grover and Baker company, and several others [64] purchased rights from Elias Howe. This gave Howe almost absolute control of the sewing-machine business as these companies agreed to his royalty terms of $25 for every machine sold. In an attempt to improve his own machine, Howe was almost immediately caught up in another series of legal battles in which he was the defendant; the companies he had defeated were able to accuse him of infringing on patents that they owned. To compound the confusion, individual companies also were suing each other on various grounds.

Because of this situation Orlando B. Potter, president of the Grover and Baker company, advanced in 1856 the idea of a "Combination" of sewing-machine manufacturers. He pointed out how the various companies were harming themselves by continuing litigation and tried to convince Howe that all would benefit by an agreement of some kind. He proposed that Elias Howe; Wheeler, Wilson and company; I. M. Singer and company; and Grover and Baker company pool their patents covering the essential features of the machine. The three companies had started production about the same time and approved of Potter's idea; Howe opposed it as he felt that he had the most to lose by joining the "Combination." He finally consented to take part in Potter's plan if the others would agree to certain stipulations. The first requirement was that at least twenty-four manufacturers were to be licensed. The second was that, in addition to sharing equally in the profits with the three companies, Howe would receive a royalty of $5 for each machine sold in the United States and $1 for each machine exported. It has been estimated that, as a result of this agreement, Howe received at least $2,000,000 as his share of the license fees between 1856 and 1867 when his patent expired.[65]

The organization was called the Sewing-Machine Trust and/or the Sewing-Machine Combination. The important patents contributed to it were:

1. The grooved, eye-pointed needle used with a shuttle to form the lockstitch (E. Howe patent, held by E. Howe);
2. The four-motion feeding mechanism (A. B. Wilson patent, held by Wheeler and Wilson company);
3. The needle moving vertically above a horizontal work-plate (Bachelder patent), a continuous feeding device by belt or wheel (Bachelder patent), a

---

[64] These included the American Magnetic Sewing Machine Co.; A. Bartholf; Nichols and Bliss; J. A. Lerow; Woolridge, Keene, and Moore; and A. B. Howe. *New York Daily Tribune*, Sept. 3, 1853.

[65] "Who Invented the Sewing-Machine," unsigned article in *The Galaxy*, vol. 4, August 31, 1867, pp. 471–481.

yielding presser resting on the cloth (Bachelder patent), the spring or curved arm to hold the cloth by a yielding pressure (Morey and Johnson patent), the heart-shaped cam as applied to moving the needle bar (Singer patent); all these patents, held by the Singer Company.[66]

The Grover and Baker company contributed several patents of relative importance, but its most important claim for admission was the fact that Potter had promoted the idea.

The consent of all four member-parties was required before any license could be granted, and all were required to have a license—even the member companies. The fee was $15 per machine. A portion of this money was set aside to pay the cost of prosecuting infringers, Howe received his initial fee, and the rest was divided between the four parties. The advantage to the licensee was that he was required to pay only one fee. Most license applications were granted; only those manufacturing a machine specifically imitating the product of a licensed manufacturer were refused.

The "Combination's" three company members each continued to manufacture, improve, and perfect its own machine. Other than the joint control of the patents, there was no pooling of interests, and each company competed to attract purchasers to buy its particular type of machine, as did the companies who were licensed by them.

In 1860, the year Howe's patent was renewed, the general license fee was reduced from $15 to $7, and Howe's special royalty was reduced to $1 per machine. Howe remained a member until his patent ran out in 1867. The other members continued the "Combination" until 1877, when the Bachelder patent, which had been extended twice, finally expired. By that time the fundamental features of the sewing machine were no longer controlled by anyone. Open competition by the smaller manufacturers was possible, and a slight reduction in price followed. Many new companies came into being—some destined to be very short-lived.

From the beginning to the end of the "Combination" there was an army of independents, including infringers and imitators, who kept up a constant complaint against it, maintaining that its existence tended to retard the improvement of the sewing machine and that the public suffered thereby. In the period immediately following the termination of the "Combination," however, only a few improvements of any importance were made, and most of these were by the member companies.

---

[66] Singer has sometimes been credited as the inventor of the various improvements covered by the patents that the Singer company purchased and later contributed to the efforts of the "Combination."

Chapter Four

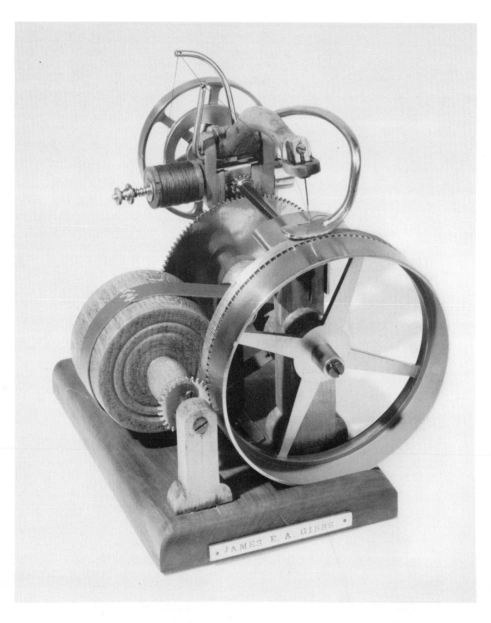

Figure 38.—Gibbs' patent model, 1857. (Smithsonian photo 45504–E.)

Less-Expensive Machines

WHILE THE "COMBINATION" was attempting to solve the problems of patent litigation, another problem faced the would-be home users of this new invention. The budget limitations of the average family caused a demand for a less-expensive machine, for this first consumer appliance was a most desirable commodity.[67]

There were many attempts to satisfy this demand, but one of the best and most successful grew out of a young man's curiosity. James E. A. Gibbs' first exposure to the sewing machine was in 1855 when, at the age of 24, he saw a simple woodcut illustration of a Grover and Baker machine. The woodcut represented only the upper part of the machine. Nothing in the illustration indicated that more than one thread was used, and none of the stitch-forming mechanism was visible. Gibbs assumed that the stitch was formed with one thread; he then proceeded to imagine a mechanism that would make a stitch with one thread. His solution was described in his own statement:

> As I was then living in a very out of the way place, far from railroads and public conveyances of all kinds, modern improvements seldom reached our locality, and not being likely to have my curiosity satisfied otherwise, I set to work to see what I could learn from the woodcut, which was not accompanied by any description. I first discovered that the needle was attached to a needle arm, and consequently could not pass entirely through the material, but must retreat through the same hole by which it entered. From this I saw that I could not make a stitch similar to handwork, but must have some other mode of fastening the thread on the underside, and among other possible methods of doing this, the chainstitch occurred to me as a likely means of accomplishing the end.
>
> I next endeavored to discover how this stitch was or could be made, and from the woodcut I saw that the driving shaft which had the driving wheel on the outer end, passed along under the cloth plate of the machine. I knew that the mechanism which made the stitch must be connected with and actuated by this driving shaft. After studying the position and relations of the needle and shaft with each other, I conceived the idea of the revolving hook on the end of the shaft, which might take hold of the thread and manipulate it into a chainstitch. My ideas were, of course, very crude and indefinite, but it will be seen that I then had the correct conception of the invention afterwards embodied in my machine.[68]

Gibbs had no immediate interest in the sewing machine other than to satisfy his curiosity. He did not think of it again until January 1856 when he was visiting his father in Rockbridge County, Virginia. While in a tailor's shop there, he happened to see a

---

[67] *Scientific American* (Jan. 29, 1859), vol. 14, no. 21, p. 165. In a description of the new Willcox and Gibbs sewing machine the following observation is made: "It is astonishing how, in a few years, the sewing machine has made such strides in popular favor, and become, from being a mechanical wonder, a household necessity and extensive object of manufacture. While the higher priced varieties have such a large sale, it is no wonder that the cheaper ones sell in such tremendous quantities, and that our inventors are always trying to produce something new and cheap."

[68] Op. cit. (footnote 53), pp. 129–131.

Figure 39.—ONE OF THE FIRST COMMERCIAL MACHINES produced by the Willcox & Gibbs Sewing Machine Co. in 1857, this machine bears no serial number, although the name "James E. A. Gibbs" is inscribed in two places on the cloth plate. It was used as the patent model for Gibbs' improvement on his 1857 patent issued the following year on August 10, 1858. (Smithsonian photo P. 6393.)

Singer machine. Gibbs was very much impressed, but thought the machine entirely too heavy, complicated, and cumbersome, and the price exorbitant. It was then that he recalled the machine he had devised. Remembering how simple it was, he decided to work in earnest to produce a less-expensive type of sewing machine.

Gibbs had little time to spend on this invention since his family was dependent upon him for support, but he managed to find time at night and during inclement weather. In contemporary references, Gibbs is referred to as a farmer, but since he is also reported to have had employers, it may be surmised that he was a farmhand. In any event, his decision to try to produce a less-expensive sewing machine suffered from a lack of proper tools and adequate materials. Most of the machine had to be constructed of wood, and he was forced to make his own needles. By the end of April 1856, however, his model was sufficiently completed to arouse the interest of his employers, who agreed to furnish the money necessary to patent the machine.

Gibbs went to Washington, where he examined sewing-machine models in the Patent Office and other machines then on the market. Completing his investigations, Gibbs made a trip to Philadelphia and showed his invention to a builder of models of new inventions, James Willcox. Much impressed with the machine, Willcox arranged for Gibbs to work with his son, Charles Willcox, in a small room in the rear of his shop. After taking out two minor patents (on December 16, 1856, and January 20, 1857), Gibbs obtained his important one, U.S. patent No. 17,427 on June 2, 1857 (fig. 38). His association with Charles Willcox led to the formation of the Willcox & Gibbs Sewing Machine Company, and they began manufacturing chainstitch machines in 1857 (fig. 39). The machine used a straight needle

Figure 40.—A DOLPHIN sewing machine based on Clark's patent of 1858. This design was first used by T. J. W. Robertson in 1855, but in his patent issued on May 22 of that year no claim was made for the machine design, only for the chainstitch mechanism. The same style was used by D. W. Clark in several of his chainstitch patents, but he also made no claim for the design, stating that the machine "may be made in any desired ornamental form." The dolphin-style machines are all chainstitch models of solid brass, originally gilt. Although only about five inches long, they are full-size machines using a full-size needle. (Smithsonian photo 45505.)

to make a chainstitch. At the forward end of the main shaft was a hook which, as it rotated, carried the loop of needle-thread, elongated and held it expanded while the feed moved the cloth until the needle at the next stroke descended through the loop so held. When the needle descended through the first loop, the point of the hook was again in position to catch the second loop, at which time the first loop was cast off and the second loop drawn through it, the first loop having been drawn up against the lower edge of the cloth to form a chain.

A Gibbs sewing machine, on a simple iron-frame stand with treadle, sold for approximately $50 in the late 1850s,[69] while a Wheeler and Wilson[70] machine or a Grover and Baker[71] with the same type of stand sold for approximately $100. After the introduction of the Gibbs machine, the Singer company[72] brought out a light family machine in 1858 that was also first sold for $100. It was then reduced to $50, but it was not popular because it was too light (see discussion of Singer machines, pp. 34–35). In 1859, Singer brought out its second, more successful family machine, which sold for $75.

Like the other companies licensed by the "Combination," Willcox and Gibbs company paid a royalty for the use of the patents it held. Although the Willcox and Gibbs machine was a single-thread chainstitch machine and the company held the Gibbs patents, the company was required to be licensed to use the basic feed, vertical needle, and other related

---

[69] *Scientific American*, vol. 15, no. 21 (January 29, 1859), p. 165, and Willcox and Gibbs advertising brochure, 1864.
[70] *Scientific American*, vol. 12, no. 8 (November 1, 1856), p. 62.
[71] Ibid., vol. 1, no. 19 (November 5, 1859), p. 303.

[72] I. M. Singer & Co.'s Gazette, vol. 5, no. 4 (March 1, 1859), p. 4, and a brochure, *Singer's New Family Sewing Machine* (in Singer Manufacturing Company, Historic Archives).

patents held by the "Sewing-Machine Combination."

With the approach of the Civil War, Gibbs returned to Virginia. Poor health prevented him from taking an active part in the war, but he worked throughout the conflict in a factory processing saltpeter for gunpowder. Afterward, Gibbs returned to Philadelphia and found that Willcox had faithfully protected his sewing-machine interests during his long absence. The firm prospered, and Gibbs finally retired to Virginia a wealthy man. Interestingly, Gibbs named the Virginia village to which he returned in later life "Raphine"—derived, somewhat incorrectly, from the Greek word "to sew."

The Willcox & Gibbs Sewing Machine Company is one of the few old companies still in existence. It discontinued making and selling family-style machines many years ago and directed its energies toward specialized commercial sewing machines, many of which are based on the original chainstitch principle.

There was also an ever-increasing number of other patentees and manufacturers who, in the late 1850s and 1860s, attempted to produce a sewing machine that would circumvent both the "Combination" and the high cost of manufacturing a more complicated type of machine. Some of the more interesting of these are pictured and described in figures 40 through 54.

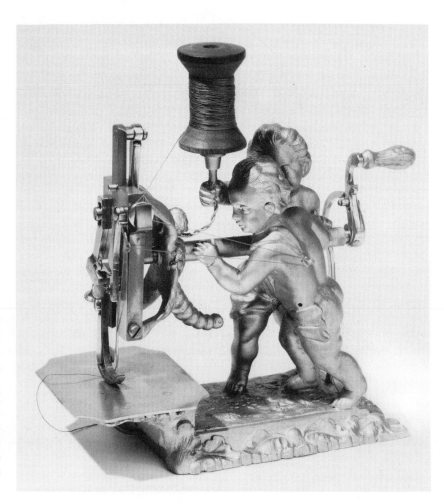

Figure 41.—THE CHERUB sewing machine was another Robertson first which was adopted by Clark. Robertson's patent of October 20, 1857, once again makes no claim for the design; neither does Clark's patent of January 5, 1858, illustrated here. The machine is approximately the same size as the dolphin and is made in the same manner and of the same materials. Two cherubs form the main support, one also supporting the spool and leashing a dragonfly which backs the needle mechanism. (Smithsonian photo 45504–D.)

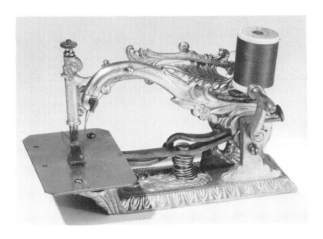

Figure 42.—The foliage sewing machine originated with D. W. Clark. Once again he did not include the design in his June 8, 1858, patent, which was aimed at improving the feeding mechanism. Like most hand-turned models, these required a clamp to fasten them to the table when in operation. (Smithsonian photo 45504–C.)

Figure 44.—The horse sewing machine is among the most unusual of the patents issued for mechanical improvements. Although James Perry, the patentee, made several claims for the looper, feeder, and tension, he made no mention of the unusual design of the machine, for which a patent was issued on November 23, 1858. Although it was probably one of a kind, the horse machine illustrates the extent to which the inventor's mind struggled for original design. (Smithsonian photo 45505–C.)

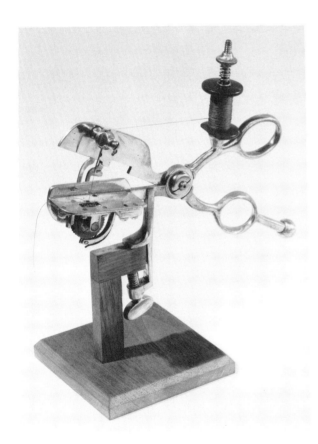

Figure 43.—The sewing shears was another popular machine of unusual style. Some models were designed to both cut and sew, but most derived their names from the method of motivating power. The earliest example of the sewing-shears machine was invented by Joseph Hendrick, who stated in his patent that he was attempting to produce "a simple, portable, cheap, and efficient machine." His patent model of October 5, 1858, is illustrated. (Smithsonian photo 45504–F.)

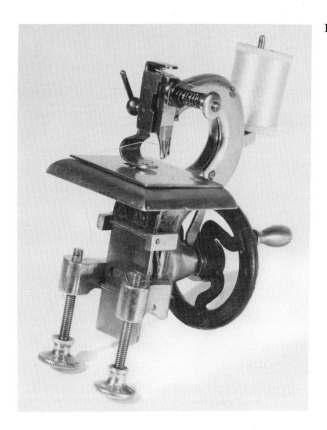

Figure 45.—MANY INVENTORS attempted to cut the cost of manufacturing a complicated machine. One of these was Albert H. Hook, whose machine is only about four inches high and two inches wide. His patent, granted November 30, 1858, simplified the construction and arrangement of the various parts. Although Hook used a barbed needle reminiscent of the one used by Thimonnier, his method of forming the stitch was entirely different. The thread was passed through the necessary guides, and when the cloth was in place the needle was thrust up from below. Passing through the fabric, the needle descended, carrying with it a loop of thread. As the process was repeated, a chainstitch was formed with the enchained loop on the under side. In spite of its simple mechanism, Hook's machine was not a commercial success. (Smithsonian photo 45505–D.)

Figure 46.—IN ADDITION TO MECHANICAL PATENTS, a number of design patents were also issued for sewing machines. These fall into a separate series in the Patent Office's numerical records. This unusual example featured two semidraped female figures holding the spool of thread, a mermaid holding the needle, a serpent which served as the presser foot, and a heart-shaped baster plate. The design was patented by W. N. Brown, October 25, 1859, but no examples other than the patent model are known to have been made. (Smithsonian photo 45504–A.)

Figure 47.—THE SQUIRREL MACHINE was another interesting design patent. S. B. Ellithorp had received a mechanical patent for a two-thread, stationary-bobbin machine on August 26, 1857. That same month he published a picture of his machine, shown here as republished in the *Sewing Machine News*, vol. 7, no. 11, November 1885. The machine was designed in the shape of "the ordinary gray squirrel so common throughout this country—an animal that is selected as a type of provident care and forethought, for its habits of frugality and for making provision for seasons of scarcity and want in times of plenty—and the different parts of the animal are each put to a useful purpose; the moving power being placed within its body, the needle stock through its head, one of its fore feet serving to guide the thread, and the other to hold down the cloth while being sewed, and the tip of its tail forming a support to the spool from which the thread is supplied."

Although the design patent was not secured until June 7, 1859, the inventor was reported to have been perfecting his machine for manufacture in 1857. Ellithorp planned "to place them in market at a price that will permit families and individuals that have heretofore been deterred from purchasing a machine by the excessive and exorbitant price charged for those now in use, to possess one." Patent rights were sold under the name of Ellithorp & Fox, but the machine was never manufactured on a large scale, if at all. No squirrel machines are known to have survived. (Smithsonian photo 53112.)

Figure 48.—HEYER'S POCKET SEWING MACHINE patent model, November 17, 1863. This patent model is one piece, and measures about two inches in height and two inches in length. It will stitch—but only coarse, loosely woven fabrics. As can be expected, a great deal of manual dexterity is required to compensate for the omission of mechanical parts. Heyer advertised patent rights for sale, but evidence of manufactured machines of this type has yet to be discovered. (Smithsonian photo 18115–D[a].)

Figure 49.—HEYER'S MACHINE as illustrated in *Scientific American*, July 30, 1864. The smallest and most original of all the attempts to simplify machine sewing, Heyer's machine, which made a chain-stitch, was constructed of a single strip of metal. The *Scientific American* stated: "It is simply a steel spring ingeniously bent and arranged and it is said to sew small articles very well. The whole affair can easily be carried in the coat pocket."

One method of operation, vibrating with the finger, was illustrated. The machine could be operated also by holding it in the hand and pressuring it between two fingers. Cloth was inserted at c, and the prongs of the spring feed f carried it along after each stitch. It was stated that the needle could be cut from the same strip of metal, but it was advised also that the needle could be made as a separate piece and attached. (Smithsonian photo 48221.)

Figure 50.—ALTHOUGH BEAN'S AND RODGERS' running-stitch machines, the second and fourth U.S. sewing-machine patents, experienced little commercial success, small manufactured machines based on Aaron Palmer's patent of May 13, 1862, were popular in the 1860s. The patent model above is a small brass implement with crimping gears that forced the fabric onto an ordinary sewing needle. The full needle was then removed from its position, and the thread was pulled through the fabric by hand. (Smithsonian photo 45524.)

THE FAIRY SEWING-MACHINE. A HOLIDAY GIFT FOR THE WORK-TABLE

Figure 51.—ONE OF THE EARLY COMMERCIAL MANUFACTURERS of the Palmer patent was Madame Demorest, a New York dressmaker. She advertised her Fairy sewing machine in *Godey's Lady's Book*, vol. 66, 1863, and stated: "In the first place it will attract attention from its diminutive, fairy-like size, and with the same ease with which it can be carried, an important matter to a seamstress or dressmaker employed from house to house . . . . What no other sewing machine attempts to do, it runs, and does not stitch, it sews the more delicate materials an ordinary sewing machine cuts or draws . . . ." (Smithsonian photo 43690.)

Figure 52.—THE FAIRY SEWING MACHINE sold for five dollars and was adequate for its advertised purpose, sewing or running very lightweight fabrics. The machine was marked with the Palmer patent, the date May 13, 1862, and the name "Mme. Demorest."

A machine identical to the Fairy, but bearing both Palmer patent dates, May 13, 1862, and June 19, 1863, and the name "Gold Medal," was manufactured by a less-scrupulous company. This machine was advertised as follows: "A first class sewing machine, handsomely ornamented, with all working parts silver plated. Put up in a highly polished mahogany case, packed ready for shipment. Price $10.00. This machine uses a common sewing needle, is very simple. A child can operate it. Cash with order." Some buyers felt they were swindled, as they had expected a heavy-duty machine, but no recourse could be taken against the advertiser. Another similar machine was also manufactured under the name "Little Gem." (Smithsonian photo 45525.)

Shaw & Clark

Figures 53 and 54.—Running-stitch machines were also attempted by several other inventors. Shaw & Clark, manufacturers of chainstitch machines, patented this running-stitch machine on April 21, 1863. From the appearance of the patent model, it was already in commercial production. On May 26, 1863, John D. Dale also received a patent for an improvement related to the method of holding the needle and regulating the stitches in a running-stitch machine. Dale's patent model was a commercial machine.

John Heberling patented several improvements in 1878 and 1880. His machine, which was a little larger and in appearance resembled a more conventional type of sewing machine, was a commercial success. (Shaw & Clark: Smithsonian photo P. 6395; Dale: Smithsonian photo P. 6394.)

Dale

Appendixes

I. Notes on the Development and Commercial Use of the Sewing Machine

INTRODUCTION

While researching the history of the invention and the development of the sewing machine, many items of related interest concerning the machine's economic value came to light. The manufacture of the machines was in itself a boost to the economy of the emerging "industrial United States," as was the production of attachments for specialized stitching and the need for new types of needles and thread. Moreover, the machine's ability to speed up production permitted it to permeate the entire field of products manufactured by any type of stitching, from umbrellas to tents. Since this aspect of the story was not completed for this study, no attempt will be made to include any definitive statements on the economic importance of the sewing machine at home or abroad. This related information is of sufficient interest, however, to warrant inclusion in this first Appendix. Perhaps these notes will suggest areas of future research for students of American technology.

READY-MADE CLOTHING

Whether of the expensive or the inexpensive type, the sewing machine was much more than a popular household appliance. Its introduction had far-reaching effects on many different types of manufacturing establishments as well as on the export trade. The newly developing ready-made clothing industry was not only in a state of development to welcome the new machine but also was, in all probability, responsible for its immediate practical application and success.

Until the early part of the second quarter of the 19th century, the ready-made clothing trade in the United States was confined almost entirely to furnishing the clothing required by sailors about to ship out to sea. The stores that kept these supplies were usually in the neighborhood of wharf areas. But other than the needs of these seamen, there was little market for ready-made goods. Out of necessity many of the families in the early years in this country had made their own clothing. As wealth was acquired and taste could be cultivated, professional seamstresses and tailors were in increasing demand, moved into the cities and towns, and even visited the smaller villages for as long as their services were needed. At the same time a related trade was also growing in the cities, especially in New York City, that of dealing in second-hand clothing. Industrious persons bought up old clothes, cleaned, repaired and refinished them, and sold the clothing to immigrants and transients who wished to avoid the high cost of new custom-made clothing.

The repairing of this second-hand clothing led to the purchase of cheap cloth at auction—"half-burnt," "wet-goods," and other damaged yardage. When in excess of the repairing needs, this fabric was made into garments and sold with the second-hand items. Many visitors who passed through New York City were found to be potential buyers of this merchandise if a better class of ready-made clothes was made available. Manufacture began to increase. Tailors of the city began to keep an assortment of finished garments on hand. When visitors bought these, they were also very likely to buy additional garments for resale at home. The latter led to the establishment of the wholesale garment-manufacturing industry in New York about 1834–35.

Most of the ready-made clothing establishments were small operations, not large factories. Large quantities of cloth were purchased; cutting was done in multiple layers with tailor's shears. Since many seamstresses were needed, the garments were farmed out to the girls in their homes. The manufacture of garments in quantity meant that the profit on each garment was larger than a tailor could make on a single custom-made item. The appeal of increased profits influenced many to enter the new industry and, due to the ensuing competition, the retail cost of each garment was lowered. Just as the new businesses were getting underway, the Panic of 1837 ruined most of them. But the lower cost and the convenience of ready-made clothing had left its mark. Not only was the garment-manufacturing business re-established soon after the Panic had subsided, but by 1841 the value of clothing sold at wholesale in New York was estimated at $2,500,000 and by 1850—a year before sewing machines were manufactured in any quantity—there were 4,278 clothing manufacturing establishments in the United States. Beside New York City, Cincinnati was also one of the important ready-made clothing centers. In 1850 the value of its products amounted to $4,427,500 and in 1860 to $6,381,190. Boston was another important center with a ready-made clothing production of $4,567,749 in 1860. Philadelphia, Baltimore, Louisville, and St. Louis all had a large wholesale clothing trade by 1860. Here was the ready market for a practical sewing machine.[73]

Clothing establishments grew and began to have agencies in small towns and the sewing work was distributed throughout the countryside. The new, competing sewing-machine companies were willing to deliver a machine for a small sum and to allow the buyer to pay a dollar or two a month until the full amount of the sale was paid. This was an extension of the hire-purchase plan (buying on credit) initiated by Clark of the Singer Company. The home seamstresses were eager to buy, for they were able to produce more piecework with a sewing machine and therefore earn more money. An example of the effect that the sewing machine had on the stitching time required was interestingly established through a series of experiments conducted by the Wheeler and Wilson company. Four hand sewers and four sewing-machine operators were used to provide the average figures in this comparative time study, the results of which were published in 1861;[74]

NUMBER OF STITCHES PER MINUTE

	By Hand	By Machine
Patent leather, fine stitching............	7	175
Binding hats...........	33	374
Stitching vamped shoes.	10	210
Stitching fine linen.....	23	640
Stitching fine silk......	30	550

TIME FOR GARMENTS STITCHED

	By Hand	By Machine
Frock coats......	16 hrs. 35 min.	2 hrs. 38 min.
Satin vests.......	7 hrs. 19 min.	1 hr. 14 min.
Summer pants...	2 hrs. 50 min.	0 hr. 38 min.
Calico dress......	6 hrs. 37 min.	0 hr. 57 min.
Plain apron......	1 hr. 26 min.	0 hr. 9 min.
Gentlemen's shirts..........	14 hrs. 26 min.	1 hr. 16 min.

The factory manufacturer, with the sewing work done at the factory, was also developing. In 1860, Oliver F. Winchester, a shirt manufacturer of New Haven, Connecticut, stated that his factory turned out 800 dozen shirts per week, using 400 sewing machines and operators to do the work of 2,000 hand sewers. The price for hand sewing was then $3 per week, which made labor costs $6000 per week. The 400 machine operators received $4 per week, making the labor cost $1600 per week. Allowing $150 as the cost of each machine, the sewing machines more than paid for themselves in less than 14 weeks, increased the operators pay by $1 a week, and lowered the retail cost of the item.[75] The greatest savings of time, which was as much as fifty percent, was in the manufacture of light goods—such items as shirts, aprons, and calico dresses. The Commissioner of Patents weighed the monetary effect that this or any invention had on the economy against the monetary gain received by the patentee. When he found that the patentee had not been fairly compensated, he had the authority to grant a seven-year extension to the patent.[76]

---

[73] *Eighth Census, 1860, Manufactures, Clothing* (United States Census Office, published Government Printing Office: Washington, D.C., 1865).

[74] *Eighty Years of Progress of the United States* (New York, 1861), vol. 2, pp. 413–429.

[75] GEORGE GIFFORD, "Argument of [George] Gifford in Favor of the Howe Application for Extension of Patent" (New York: United States Patent Office, 1860).

[76] Op. cit. (footnote 34).

The sewing machine also contributed to the popularity of certain fashions. Ready-made cloaks for women were a business of a few years' standing when the sewing machine was adopted for their manufacture in 1853. Machine sewing reduced the cost of constructing the garment by about eighty percent, thereby decreasing its price and increasing its popularity. In New York City alone, the value of the "cloak and mantilla" manufacture in 1860 was $618,400.[77] Crinolines and hoopskirts were easier to stitch by machine than by hand, and these items had a spirited period of popularity due to the introduction of the sewing machine. Braiding, pleating, and tucking adorned many costume items because they could be produced by machine with ease and rapidity.

In addition to using the sewing machine for the manufacture of shirts, collars, and related men's furnishings, the machine was also used in the production of men's and boy's suits and reportedly gave "a vast impetus to the trade."[78] The Army, however, was not quite convinced of the sewing machine's practical adaptation to its needs. Although a sewing machine was purchased for the Philadelphia Quartermaster Depot as early as 1851, they had only six by 1860. On March 31, 1859, General Jesup of the Philadelphia Depot wrote to a Nechard & Company stating that the machine sewing had been tried but was not used for clothing, only for stitching caps and chevrons. In another letter, on the same day, to "Messers Hebrard & Co., Louisiana Steam Clothing Factory, N. Orleans," Jesup states: "Machine sewing has been tried with us, and though it meets the requirements of a populous and civilized life, it has been found not to answer for the hard wear and tear and limited means of our frontier service. Particular attention has been paid to this subject, and we have abandoned the use of machines for coats, jackets and trousers, etc and use them on caps and bands that are not exposed to much hard usage. . . ."[79] At this period prior to the Civil War, the Army manufactured its own clothing. As the demands of war increased, more and more of the Army's clothing supplies were furnished on open contract—with no

Figure 55.—BLAKE'S LEATHER-STITCHING MACHINE patent model of July 6, 1858; the inventor claimed the arrangement of the mechanism used and an auxiliary arm capable of entering the shoe, which enabled the outer sole to be stitched both to the inner sole and to the upper part of the shoe. (Smithsonian photo 50361.)

specifications as to stitching.[80] Machine stitching, in fact, is found in most of the Civil War uniforms. One of the problems that most probably affected the durability of the machine stitching in the 1850s was the sewing thread, a problem that was not solved until the 1860s and which is discussed later under "thread for the machine."

SHOE MANUFACTURE

Another industry that was aided by the new invention was that of shoe manufacture. Although the earliest sewing-machine patents in the United States reflect the inventors' efforts to solve the difficult task of leather stitching, and, although machines were used to a limited extent in stitching some parts of

---

[77] *Eighth Census, 1860, Manufactures* (United States Census Office, published Government Printing Office: Washington, D.C., 1865), "Women's Ready-Made Clothing," p. 83.

[78] Ibid., p. 64.

[79] National Archives, Record Group 92, Office of the Quartermaster General, Clothing Book, Letters Sent, volume 17.

---

[80] The author wishes to acknowledge the valuable help of Mr. Donald Kloster of the Smithsonian Institution's Division of Military History for the preceding four references and related information.

Figure 56.—HARRIS' patent thread cutter, 1872. (Smithsonian photo P-6397.)

Figure 57.—WEST's patent thread cutter, 1874. (Smithsonian photo P-63100.)

Figure 58.—KARR's patent needle threader, 1871. (Smithsonian photo P-63101.)

the shoe in the early and mid-1850s, it was not until 1858 that a machine was invented that could stitch the sole to the inner sole and to the upper part of the shoe. This was the invention of Lyman R. Blake and was patented by him on July 8, 1858; the patent model is shown in figure 55. Blake formed a chainstitch by using a hooked needle, which descended from above, to draw a thread through the supporting arm. Serving as the machine's bedplate, the arm was shaped to accommodate the stitching of all the parts of the shoe.

The increased number of shoes required by the Army during the Civil War spurred the use of the sewing machine in their manufacture. The first "machine sewed bootees" were purchased by the Army in 1861. Inventors continued their efforts; the most prominent of these was Gordon McKay, who worked on an improvement of the Blake machine with Robert Mathies in 1862 and then with Blake in 1864. Reportedly, the Government at first preferred the machine-stitched shoes as they lasted eight times longer than those stitched by hand; during the war the Army purchased 473,000 pairs, but in 1871 the Quartermaster General wrote:

> No complaints regarding the quality of these shoes were received up to February 1867 when a Board of Survey, which convened at Hart's Island, New York Harbor reported upon the inferior quality of certain machine sewed bootees of the McKay patent, issued to the enlisted men at that post. The acting Quartermaster General, Col. D. H. Rucker, April 10, 1867, addressed a letter to all the officers in charge of depots, with instructions not to issue any more of the shoes in question, but to report to this office the quantity remaining in store. From these reports it appears that there were in store at that time 362,012 pairs M. S. Bootees, all of which were ordered to be, and have since been sold at public auction.[81]

The exact complaint against the shoes was not recorded. Possibly the entire shoe was stitched by machine. It was found that although machine-stitched shoes were more durable in some respects and the upper parts of most shoes continued to be machine stitched, pegged soles for the more durable varieties remained the fashion for a decade or more, as did custom hand-stitched shoes for those who could afford them.

OTHER USES

The use of sewing machines in all types of manufacturing that required stitching of any type continued to grow each year. While the principal purpose for which they were utilized continued to be the manufacture of clothing items, by the year 1900 they were also used for awnings, tents, and sails; cloth bags; bookbinding and related book manufacture; flags and banners; pocketbooks, trunks, and valises; saddlery and harnesses; mattresses; umbrellas; linen and rubber belting and hose; to the aggregate sum of nearly a billion dollars—$979,988,413.[82]

SEWING-MACHINE ATTACHMENTS

The growing popularity of the sewing machine offered still another boost to the economy—the development of many minor, related manufacturing

---

[81] Letter of Nov. 4, 1871, to Col. Theo. A. Dodge, USA (Ret.), Boston, from Quartermaster General M. C. Meigs, in the National Archives, Record Group 92, Quartermaster General's Office, Letters Sent, Clothing Supplies, 1871.

[82] *Twelfth Census of the United States, 1900*, vol. 10, *Manufactures*, Part 4, Special Reports on Selected Industries (United States Census Office, Washington, D.C., 1902).

Figure 59.—SHANK's patent bobbin winder, 1870. (Smithsonian photo P-6398.)

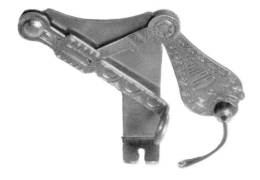

Figure 62.—ROSE's patent embroiderer, 1881. (Smithsonian photo P-6399.)

Figure 60.—SWEET's patent binder, 1853. (Smithsonian photo P-6396.)

Figure 63.—HARRIS' patent buttonhole attachment, 1882. (Smithsonian photo P-63103.)

Figure 61.—SPOUL's patent braid guide, 1871. (Smithsonian photo P-63102.)

industries. The repetitive need for machine needles, the development of various types of attachments to simplify the many sewing tasks, and the ever-increasing need for more and better sewing thread—the sewing machine consumed from two to five times as much thread as stitching by hand—created new manufacturing establishments and new jobs.

The method of manufacturing machine needles did not differ appreciably from the method used in making the common sewing needle, but the latter had never become an important permanent industry in the United States. Since the manufacture of practical sewing machines was essentially an American development and the eye-pointed needle a vital component of the machine, it followed that the manufacture of needles would also develop here. Although such a manufacture was established in 1852,[83] foreign imports still supplied much of the

---

[83] CHARLES M. KARCH, "Needles: Historical and Descriptive," in *Twelfth Census of the United States, 1900*, vol. 10, *Manufactures*, Part 4, Special Reports on Selected Industries (United States Census Office: Washington, D.C., 1902), pp. 429–432.

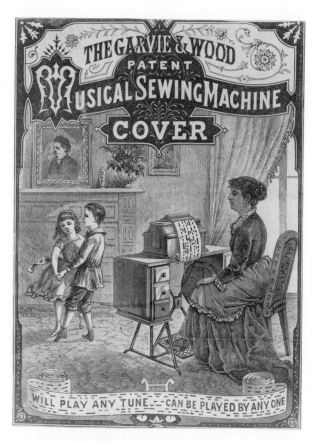

Figure 64.—THE TREADLE OF THE MACHINE was also used to help create music. George D. Garvie and George Wood received patent 267,874, Nov. 21, 1882, for "a cover for a sewing machine provided with a musical instrument and means for transmitting motion from the shaft of the sewing machine to the operating parts of the musical instrument." Although no patent model was submitted by the inventors, the "Musical Sewing Machine Cover" was offered for sale as early as October 1882, as shown by this advertisement that appeared in *The Sewing Machine News* that month. (Smithsonian photo 57983.)

Figure 65.—THIS FANNING ATTACHMENT was commercially available from James Morrison & Co. in the early 1870s; it sold for one dollar as stated in the advertising brochure from which this engraving was copied. Other inventors also patented similar implements. (Smithsonian photo 45513.)

need in the 1870s. As more highly specialized stitching machines were developed, an ever-increasing variety of needles was required, and the industry grew.

Soon after the sewing machine was commercially successful, special attachments for it were invented and manufactured. These ranged from the simplest devices for cutting thread to complicated ones for making buttonholes (see figs. 56 through 66).

The first patent for an attachment was issued in 1853 to Harry Sweet for a binder, used to stitch a special binding edge to the fabric. Other related attachments followed; among these were the hemmer which was similar to the binder, but turned the edge of the same piece of fabric to itself as the stitching was performed. Guides for stitching braid in any pattern, as directed by the movement of the goods below, were also developed; this was followed by the embroiderer, an elaborate form of braider. The first machine to stitch buttonholes was patented in 1854 and the first buttonhole attachment in 1856, but the latter was not practical until improvements were made in the late 1860s. Special devices for refilling the bobbins were invented and patented as early as 1862, and the popularity of tucked and ruffled garments inspired inventors to provide sewing-machine attachments for these purposes also. To keep the seamstress cool, C. D. Stewart patented an attachment for fanning the operator by an action derived from the treadle (fig. 65). While electric

Figure 66.—Jones "electro motor" patent model of 1871 on a Bartlett sewing machine. (Smithsonian photo P-63104.)

sewing machines did not become common until the 20th century, several 19th-century inventors considered the possibility of attaching a type of motor to the machine. One was the 1871 patent of Solomon Jones, who added an "electro motor" to an 1865 Bartlett machine (fig. 66). The attachments that were developed during the latter part of the 19th century numbered in the thousands; many of these were superfluous. Most of the basic ones in use today were developed by the 1880s and remain almost unchanged. Even the recently popular home zigzag machine, an outgrowth of the buttonhole machine, was in commercial use by the 1870s.

Sewing-machine improvements have been made from time to time. Like other mechanical items the machine has become increasingly automatic, but the basic principles remain the same. One of the more recent developments, patented [84] in 1933 by Valentine Naftali et al., is for a manufacturing machine that imitates hand stitching. This machine uses a two-pointed "floating needle" that is passed completely through the fabric—the very idea that was attempted

---

[84] U.S. patent 1,931,447, issued to Valentine Naftali, Henry Naftali, and Rudolph Naftali, Oct. 17, 1933. The Naftali machines are manufactured by the American Machine and Foundry Company and are called AMF Stitching Machines.

Figure 67.—Six-cord cabled thread.

over one hundred years ago. The machine is currently used by commercial manufacturers to produce decorative edge-stitching that very closely resembles hand stitching.

THREAD FOR THE MACHINE

The need for a good thread durable enough to withstand the action of machine stitching first created a problem and ultimately another new industry in this country. When the sewing machine was first developed the inventors necessarily had to use the sewing thread that was available. But, although the contemporary thread was quite suitable for hand sewing, it did not lend itself to the requirements of the machine. Cotton thread, then more commonly a three-ply variety, had a glazed finish and was wiry. Silk thread frequently broke owing to abrasion at the needle eye. For the most part linen thread was too coarse, or the fine variety was too expensive. All of the thread had imperfections that went unnoticed in the hands of a seamstress, but caused havoc in a machine.

Quality silk thread that would withstand the rigors of machine stitching could be produced, but it was quite expensive also. A new type of inexpensive thread was needed; the obvious answer lay in improving the cotton thread.[85]

In addition to the popular three-ply variety, cotton thread was also made by twisting together either two single yarns or more than three yarns. Increasing the number of yarns produced a more cylindrical thread. The earliest record of a six-ply cotton thread was about 1840.[86] And in 1850 C. E. Bennett of Portsmouth, New Hampshire, received a gold medal for superior six-cord, or six-ply, spool cotton at the Fair of the American Institute. But the thread was still wiry and far from satisfactory. By the mid-1860s the demonstrated need for thread manufacturers in America brought George A. Clark and William Clark, third generation cotton-thread manufacturers of Paisley, Scotland, to Newark, New Jersey, where they built a large mill. George Clark decided that a thread having both a softer finish and a different construction was needed. He produced a six-cord cabled thread, made up of three two-ply yarns (see fig. 67). The thread was called "Clark's 'Our New Thread,'" which was later shortened to O.N.T. The basic machine-thread problem was solved. When other manufacturers used the six-cord cabled construction they referred to their thread as "Best Six-Cord"[87] or "Superior Six-Cord"[88] to distinguish it from the earlier variety made up of six single yarns in a simple twist. Another new side industry of the sewing machine was successfully established.

MANUFACTURE AND EXPORT, TO 1900

Sewing machines were a commodity in themselves, both at home and abroad. In 1850, there were no establishments exclusively devoted to the manufacture of sewing machines, the few constructed were made in small machine shops. The industry, however, experienced a very rapid growth during the next ten years. By 1860 there were 74 factories in 12 States,[89] mainly in the East and Midwest,[90] producing over 111,000 sewing machines a year. In addition, there were 14 factories that produced sewing-machine cases and attachments. The yearly value of these products was approximately four and a half million dollars, of which the amount exported in 1861 was $61,000. Although the number of sewing-machine factories dropped from 74 in 1860 to 69 by 1870, the value of the machines produced increased to almost sixteen million dollars.

The number of sewing-machine companies fluctuated greatly from year to year as many attempted to enter this new field of manufacture. Some were not able to make a commercial success of their products. The Civil War did not seem to be an important factor in the number of companies in business in the North. Although one manufacturer ceased operations in Richmond, Virginia, and a Vermont firm converted to arms manufacture, several companies began operations during the war years. Of the 69 firms in business in 1870, only part had been in business since 1860 or before; some were quite new as a result of the expiration of the Howe patent renewal in 1867.

Probably due to the termination of many of the major patents, there were 124 factories in 1880, but the yearly product value remained at sixteen million dollars. The 1890 census reports only 66 factories with a yearly production of a little less than the earlier decade. But by 1900, the yearly production of a like number of factories had reached a value of over twenty-one million, of which four and a half million dollars worth were exported annually. The total value of American sewing machines exported from 1860 to 1900 was approximately ninety million dollars. The manufacture of sewing machines made a significant contribution to the economic development of 19th-century America.

---

[85] See Appendix VII, p. 216, "A Brief History of Cotton Thread."

[86] *The Story of Cotton Thread* (New York, The Spool Cotton Company, 1933).

[87] J. and P. Coats spool cotton.

[88] Willimantic spool cotton.

[89] New Hampshire, Vermont, Massachusetts, Rhode Island, Connecticut, New York, Pennsylvania, Delaware, Ohio, Indiana, Illinois, Kentucky. *Eighth Census, 1860, Manufactures* (United States Census Office, published by Government Printing Office: Washington, D.C., 1865.)

[90] Sewing-machine manufacture in the South was just beginning to blossom when it was curtailed by the outbreak of the Civil War. See Lester sewing machine, figure 109 on page 104.

II. American Sewing-Machine Companies of the 19th Century

During the second half of the 19th century there were more than two hundred sewing-machine companies in the United States. A few of the companies manufactured commercial-type machines for factories, which were increasing in number and type with each decade; however, most of the companies were primarily concerned with the manufacture of sewing machines for the home. A representative number of these family machines with information concerning the company and dating by serial numbers are pictured in figures 68 through 132. A great many of the companies were licensed by the "Combination," but not all. Some companies were constructing machines that did not infringe the patents, some companies infringed the patents but managed to avoid legal action, and many companies mushroomed into existence after the major patents expired and the "Combination" was dissolved in 1877. Of this latter group, most were short-lived. Although they were free of the royalty charges of the earlier decades, it was difficult to compete with the established companies. A few were successful. Establishing the exact dates that the companies were in existence is difficult. Their records were incomplete or have disappeared. Many of the "Combination" records were lost by fire. A summary of the existing records kept by the "Combination" of the companies paying royalty for the patents held is given in figure 37.

As will be noted in the following listing, only a small percentage of the sewing-machine companies were in business for a period of more than ten years. Of those that continued longer, all but a few dozen had disappeared by 1910. In the 1960s there were about sixty sewing-machine companies listed in *Thomas' Register of Manufacturers*, many of which were manufacturing highly specialized sewing machines for commercial work; only a few produced family- or home-sewing machines. By 1975 it was difficult to find a single family machine that was totally produced in the United States, although several companies still distributed sewing machines under their American company names. Foreign competition had increased, and the high cost of skilled labor in this country had made competition in this consumer-product field increasingly difficult. The countless varieties of American-made family sewing machines, so evident in the 19th and early 20th centuries, were no more.

Sewing Machine	Manufacturer or Company	First Made or Earliest Record	Discontinued or Last Record
Aetna	Plaver, Braunsdorf, & Co., Boston	ca. 1867	ca. 1869
	J. E. Braunsdorf & Co.	ca. 1869	ca. 1877
Akins and Felthousen	——, Ithaca, N.Y.	ca. 1855	—
Alsop	——	—	ca. 1880
American Buttonhole, Overseaming and Sewing Machine (Fig. 68)	American Buttonhole, Overseaming and Sewing Machine Co., Philadelphia, Pa.	1867	ca. 1877
Later New American (Fig. 69)	American Sewing Machine Co., Philadelphia, Pa.	ca. 1877	after 1888

Sewing Machine	Manufacturer or Company	First Made or Earliest Record	Discontinued or Last Record
American Magnetic (Fig. 70)	American Magnetic Sewing Machine Company, Ithaca, N.Y.	1853	1854
Atlantic (fig. 71)	——	1869	ca. 1870
Atwater (fig. 87)	——	1857	ca. 1866
Avery	Otis Avery, New York, N.Y.	1852	ca. 1856
Avery	Avery Manufacturing Co., Bridgeport, Conn.	1875	ca. 1887
	Avery Sewing Machine, New York, N.Y.	—	1888
A. Bartholf Manufr. Blodgett & Lerow patent 1849. *See* also	A. Bartholf, manufacturer, New York, N.Y.	ca. 1850	185–
A. Bartholf Manufr. Howe's patent, 1846 (fig. 72)	A. Bartholf, manufacturer, New York, N.Y.	1853	ca. 1856
Bartholf	A. Bartholf, manufacturer	1857	1859
	Bartholf Sewing Machine Co.	1859	ca. 1865
Bartlett (fig. 73)	Goodspeed & Wyman, Wichendon, Mass.	1866	ca. 1870
Bartlett Reversible	Bartlett Sewing Machine Co., New York, N.Y.	ca. 1870	1876
Baker	——	—	before 1880
Bartram & Fanton (fig. 74)	Bartram & Fanton Mfg. Co., Danbury, Conn.	1867	1874
Bay State	——	—	before 1880
Beckwith (fig. 75)	Barlow & Son, New York, N.Y.	1871	1872
	Beckwith Sewing Machine Co., New York, N.Y.	1872	ca. 1877
Blees	Blees Sewing Machine Co., Bordentown, N.J.	1870	1873
Blodgett & Lerow (fig. 21)	O. Phelps, Boston, Mass.	1849	1849
	Goddard, Rice & Co., Worcester, Mass.	1849	1850
(fig. 20)	A. Bartholf, manufacturer, New York, N.Y.	1849	185–
Bond	——	—	before 1880
Boston	J. F. Paul & Co., Boston, Mass.	1880	—
Later New Boston	Boston Sewing Machine Co., Boston, Mass.	1881	—
	Bi-Spool Sewing Machine Co., Boston, Mass.		
	Acme Manu. Co., Boston, Mass.	—	1888
Boudoir (fig. 76)	Daniel Harris, inventor and patentee Manufacturer—several	1857	ca. 1870
Bradford & Barber	Bradford & Barber, manufacturers, Boston, Mass.	1860	1861

Sewing Machine	Manufacturer or Company	First Made or Earliest Record	Discontinued or Last Record
Bradshaw Shuttle	George B. Sloat & Co., Philadelphia, Pa.	ca. 1859	1861
Brattleboro	Samuel Barker and Thomas White, Brattleboro, Vt.	ca. 1858	1861
Brown Rotary	Brown Rotary Shuttle Sewing Machine Co., Indianapolis, Ind.	—	1878
Bruen	Bruen Manufacturing Co., N.Y.C.	ca. 1866	ca. 1870
Buckeye	Wilson [W. G.] Sewing Machine Co., Cleveland, Ohio	ca. 1867	ca. 1876
Later New Buckeye (fig. 77) (see Wilson)			
Buell, "E. T. Lathbury's Patent"	A. B. Buell, Westmoreland, New York	ca. 1860	—
Burnet & Broderick	Burnet, Broderick and Co.	1859	ca. 1860
Butterick	The Eclipse Sewing Machine Co., Cincinnati, Ohio	1888	after 1888
Carolina	Carolina Manufacturing Co., Shelby, N.C.	1884	—
Centennial (fig. 78)	Centennial Sewing Machine Co., (see McLean and Hooper), Philadelphia, Pa.	1873	1876
Chamberlain	Woolridge, Keene and Moore, Lynn, Mass.	1853	ca. 1854
Chicago Singer	Scates, Tryber & Sweetland Mfg. Co., Chicago, Ill.	1879	1882
Later Chicago	Chicago Sewing Machine Co., Chicago, Ill.	1882	after 1888
Chicopee	Chicopee Sewing Machine Co., Chicopee Falls, Mass.	1868	ca. 1869
(see Shaw & Clark)			
Cincinnati	Queen City Sewing Machine Co., Cincinnati, Ohio		
Clark (fig. 42)	D. W. Clark, Bridgeport, Conn.	ca. 1858	after 1860
Clark's Revolving Looper [double thread] (fig. 79) (see Windsor)	Lamson, Goodnow & Yale, Windsor, Vt.	1859	1861
Clinton	Clinton Brothers, Ithaca, N.Y.	ca. 1861	ca. 1865
Companion	Thurston, Mfg. Co., Marlboro, N.H. moved to New Britain, Conn.	1882 —	— by 1888
Crown	Florence Sewing Machine Co., Florence, Mass.	1879	ca. 1888
(see Florence)			
Daisy ($5 Machine)	Daisy Sewing Machine Co., Cleveland, Ohio	ca. 1877	ca. 1883–1888

Sewing Machine	Manufacturer or Company	First Made or Earliest Record	Discontinued or Last Record

Dauntless
    Later New Dauntless
    (see Queen)
	Dauntless Mfg. Co., Norwalk, Ohio	1877	after 1882

Davis	J. A. Davis, New York, N.Y.	ca. 1860	—
Davis Vertical Feed	Davis Sewing Machine Co., Watertown, N.Y.	1869	1889
Davis Vertical Feed and Rotary Shuttle	Davis Sewing Machine Co., Dayton, Ohio	after 1886	1924
Davis Rotary (see Rotary Shuttle)			
Decker (also The Princess)	Decker Mfg. Co., Detroit, Mich.	—	before 1881
Demorest	Demorest Mfg. Co. (formerly N.Y. Sewing Machine Co.) N.Y.C.	1882	1888
	Plattsburgh, New York	1888	1908
Diamond (formerly Sigwalt)	Sigwalt Sewing Machine Co., Chicago, Ill.	1880	—
	Diamond Sewing Machine Co., Chicago, Ill.	—	in 1888
Domestic	Wm. A. Mack & Co., and N. S. C. Perkins, Norwalk, Ohio	1864	1869
Domestic	Domestic Sewing Machine Co., Norwalk, Ohio, acquired by White Sewing Machine Co., in 1924 and maintained as a subsidiary at Cleveland, Ohio	1869	*
Domestic	Domestic Sewing Machine Co., Newark, N. J. (sold to Domestic Sewing Machine Co. of Ohio)	1872	ca. 1906
Dorcas	The American Sewing Machine Co., (John P. Bowker, Agent) Boston, Mass.	1853	before 1856
Du Laney (fig. 80) Also called Little Monitor (see)			
Durgin	Charles A. Durgin, New York, N.Y.	1853	after 1855
Eclipse, formerly Cincinnati	The Eclipse Sewing Machine Co., Cincinnati, Ohio	1885	1888
Elastic Motion	Elastic Motion Sewing Machine Co., Brooklyn, N.Y.	—	out by 1888
Eldredge	Eldredge Sewing Machine Co., Chicago, Ill.	1869	—
	Eldredge Manufacturing Co., Belvidere, Ill.	1888	1890
Elliptic			
Sloat's Elliptic	George B. Sloat and Co., Philadelphia, Pa.	ca. 1858	ca. 1860
Sloat's Elliptic	Union Sewing Machine Co., Richmond, Va.	1860	1861

* Still in existence.

Sewing Machine	Manufacturer or Company	First Made or Earliest Record	Discontinued or Last Record
Elliptic	Wheeler & Wilson Mfg. Co.	1861	ca. 1867
	Elliptic Sewing Machine Co., New York, N.Y.	1867	before 1880
Empire (fig. 86)	Empire Co., Boston, Mass.	ca. 1860	ca. 1870
Empire Later Remington-Empire	Empire Sewing Machine Co., New York	1866	ca. 1870
Empress	Manufactured on order through Jerome B. Secor, Bridgeport, Conn.	1877	—
Estey	Estey Sewing Machine Co.	ca. 1880	1882
Estey, Fuller-Model	Brattleboro Sewing Machine Co., Brattleboro, Vt.	1883	out by 1888
Eureka Shuttle (fig. 81)	Eureka Sewing Machine Co., New York, N.Y.	1859	1860
Excelsior	Excelsior Sewing Machine Co., New York, N.Y.	1854	1854
Fairbanks	Belleville Mfg. Co., Belleville, Ill. Fairbanks Sewing Machine Co., Springfield, Ill.	—	out by 1888
Fairy	Jerome B. Secor, Bridgeport, Conn.	1883	—
Fairy (figs. 51, 52)	Madame Demorest, New York, N.Y.	1863	ca. 1865
Finkle, M. (fig. 82)	M. Finkle, Boston, Mass.	1856	ca. 1859
Finkle & Lyon	Finkle & Lyon Sewing Machine Co., Boston, Mass.	ca. 1859	1867
Later Victor			
First and Frost	First and Frost, New York, N.Y.	ca. 1859	ca. 1861
Florence (fig. 83) Later Crown	Florence Sewing Machine Co., Florence, Mass.	ca. 1860	after 1878
Folsom (*see* Globe and New England)	Folsom, J. G., Winchendon, Mass.	1865	ca. 1872
Fosket and Savage	Fosket and Savage, Meriden, Conn.	1858	1859
Foxboro (*see* Rotary Shuttle)			
Franklin	Franklin Sewing Machine Co., Mason Village, N.H.	1871	1871
Free	Free Sewing Machine Co., Chicago and Rockford, Ill.	1898	1958
	Free Sewing Machine Co., Los Angeles, Cal. (purchased by Janome Co. of Japan)	1958	1960
Gardner	C. R. Gardner, Detroit, Mich.	1856	—
Gardner	——, Norwalk, Ohio	1860	1864
Globe (figs. 84, 85)	J. G. Folsom, Winchendon, Mass.	1865	1869
Gold Medal (chainstitch)	A. G. Johnson & Co.	1867	1869
	Gold Medal Sewing Machine Co., Orange, Mass.	1869	1876

Sewing Machine	Manufacturer or Company	First Made or Earliest Record	Discontinued or Last Record
Gold Medal (running stitch)	——	1863	ca. 1865
Gold Hibbard	Hibbard, B. S., & Co.	1875	—
Goodbody (sewing shears)	Goodbody Sewing Machine Co., Bridgeport, Conn.	1880	ca. 1890
Goodes	Rex & Bockius, Philadelphia, Pa.	ca. 1876	before 1881
Goodrich	H. B. Goodrich, Chicago, Ill.	ca. 1880	1885
Later Foley & Williams	Foley & Williams Mfg. Co., Chicago	1885	ca. 1924
	Goodrich Sewing Machine Co., Chicago, Ill.	ca. 1924	ca. 1935
Grant Brothers (fig. 90)	Grant Bros. & Co., Philadelphia, Pa.	1867	ca. 1870
Greenman and True (fig. 91)	Greenman and True Mfg. Co., Norwich, Conn.	1859	1860
	Morse and True	1860	1861
Green Mountain	——	ca. 1860	—
Griswold Variety	L. Griswold, New York, N.Y.	ca. 1886	ca. 1890
Grover and Baker (figs. 34–36, 92)	Grover and Baker Sewing Machine Co., Boston, Mass.	1851	1875
Hancock (figs. 93, 94)	——	1868	before 1881
Heberling Running Stitch	Heberling Running Stitch Co., made by The Brown & Sharp Mfg. Co., Providence, R. I.	1880	ca. 1885
Helpmate	Williams Mfg. Co., Plattsburgh, N.Y.	1888	—
Herron's Patent (fig. 95)	——	1857	—
Higby Later Acme	Higby Sewing Machine Co., Brattleboro, Vt.	ca. 1882	after 1886
Home Home Shuttle	Johnson, Clark & Co., Orange, Mass. *See* New Home	1869	after 1876
Homestead	——	ca. 1881	—
Household	Providence Tool Co., Providence, R.I.	1873	ca. 1884
	Household Sewing Machine Co.	ca. 1885	1906
Howe (figs. 96, 97)	Howe Sewing Machine Co., New York, N.Y. (company of A. B. Howe sold to Howe Machine Co.)	1853	1873
Howe (fig. 98)	Howe Machine Co., Bridgeport, Conn.	1867	1886
Howe's Improved Patent (fig. 107)	Nichols and Bliss, Boston, Mass.	1852	1853
	J. B. Nichols & Co.	1853	1854
which became Leavitt	Nichols, Leavitt & Co., Boston, Mass.	1854	1856
N. Hunt, which became Hunt and Webster (figs. 99, 100) Later Ladd and Webster (*see*)	N. Hunt & Co., Boston, Mass. Hunt and Webster, Boston, Mass.	1853 1854	1854 1857

Sewing Machine	Manufacturer or Company	First Made or Earliest Record	Discontinued or Last Record
Illinois	Illinois Sewing Machine Co., Rockford, Ill.	1895	ca. 1897
Improved Common Sense (fig. 102)	——	ca. 1870	—
Independent Noiseless	Independent Sewing Machine Co., Binghamton, N.Y.	1873	—
Jennie June	June Mfg. Co., Chicago, Ill. Later Belvidere, Ill.	1879	1890
Jewel	Jewel Mfg. Co., Toledo, Ohio	1884	1888
Toledo "Blade"	Jewel Mfg. Co., Toledo, Ohio	1888	1890
Johnson (fig. 103)	Emery, Houghton & Co., Boston, Mass.	1856	after 1865
Keystone	Keystone Sewing Machine Co.	before 1872	ca. 1874
Kruse Automatic	Kruse Mfg. Co., N.Y.C.	1888	—
Ladd & Webster (fig. 101)	Ladd, Webster & Co., Boston, Mass.	1858	ca. 1866
Ladies Companion (fig. 115) (see Pratt's Patent)	——	1858	ca. 1858
"Lady" (fig. 104)	——	1859	—
Landfear's Patent (fig. 105)	Parkers, Snow, Brooks & Co., West Meriden, Conn.	1857	—
Langdon	L. W. Langdon	1856	—
Lathrop (fig. 106)	Lathrop Combination Sewing Machine Co.	1873	—
Leader	Leader Sewing Machine Co., Springfield, Mass.	1882	out by 1888
Leavitt (fig. 108)	Nichols, Leavitt & Co., Boston, Mass.	1855	1857
	Leavitt & Co.	1857	ca. 1865
	Leavitt Sewing Machine Co.	ca. 1865	1870
Leslie Revolving Shuttle	Leslie Sewing Machine Co., Cleveland, Ohio	1881	out by 1888
Lester (fig. 109)	J. H. Lester, Brooklyn, N.Y.	ca. 1858	early 1860
Lester Plantation	Lester Mfg. Co., Richmond, Va.	early 1860	late 1860
	Union Sewing Machine Co., Richmond, Va.	late 1860	1861
Lever Motion	Lever Motion Sewing Machine Co., N.Y.C.	—	before 1888
Little Gem	——	—	ca. 1870
Little Giant	——, Norwalk, Ohio	ca. 1860	1864
Little Monitor (not associated with Monitor)	G. L. Du Laney & Co., Brooklyn, N.Y.	ca. 1866	after 1876
Love	Love Mfg. Co., Pittsburgh, Pa., factory in Rochester, Pa.	1885	about 1889
Lyon	Lyon Sewing Machine Co.	1879	ca. 1880

Sewing Machine	Manufacturer or Company	First Made or Earliest Record	Discontinued or Last Record
Macauley	Thos. A. Macauley Mfg., New York, N.Y.	before 1879	—
Manhattan	Manhattan Sewing Machine Co.	ca. 1868	ca. 1880
McKay	McKay Sewing Machine Assoc.	1870	1876
McLean and Hooper (see Centennial)	B. W. Lacy & Co., Philadelphia, Pa.	ca. 1869	ca. 1873
Melone	Melone Sewing Machine Co., Chillicothe, Ohio	—	out by 1888
Meyers	J. M. Meyers	1859	—
Miller's Patent	——	1853	—
Minnehaha	William M. Shaw, Cincinnati, Ohio	ca. 1870	ca. 1870
Monitor (fig. 88)	Shaw & Clark Sewing Machine Co., Biddeford, Me.	1860	1864
Moore	Moore Sewing Machine Co.,	ca. 1860	—
Morey & Johnson (fig. 18)	Safford & Williams Makers, Boston, Mass.	1849	ca. 1851
Morrison	Morrison, Wilkinson & Co., Hartford, Conn.	1881	out by 1888
Mower	——, Norwalk, Ohio	ca. 1863	1864
National	Johnson, Clark & Co., Orange, Mass.	1874	—
National (also sold under distributor's name)	National Sewing Machine Co., (consolidation of the June and Eldredge Companies), Belvidere, Ill.	1890	1953
Ne Plus Ultra (fig. 110)	O. L. Reynolds Manufacturing Co., Dover, N.H.	1857	—
Nettleton & Raymond (fig. 111)	Nettleton & Raymond, Brattleboro, Vt.	ca. 1857	—
New England (figs. 112, 113)	Charles Raymond	ca. 1859	1866
	Grout & White, Orange, Mass.	1862	1863
	Clark & Barker, Orange, Mass.	1862	1865
	Wm. Grout, Winchendon, Mass.	1863	—
	J. G. Folsom, Winchendon, Mass.	1865	1865
	Clark, Orange, Ma.	1865	1867
	A. F. Johnson, Orange, Ma.	1867	1867
Newell	——	1881	—
New Fairbanks	J. H. Drew & Co.	1878	1880
	Thomas M. Cochrane Co., Belleville, Ill.	1880	—
New Home	Johnson, Clark & Co., New Home Sewing Machine Co., Orange, Mass. (in 1928 became affiliated with Free Sewing Machine Co., which see) now owned by Janome Co. of Japan	1877 1882	1882 1959–1975

Sewing Machine	Manufacturer or Company	First Made or Earliest Record	Discontinued or Last Record
New Howe	New Howe Mfg. Co., Bridgeport, Conn.	1887	ca. 1889
New York	——, New York, N.Y.	ca. 1855	ca. 1855
New York Shuttle	N. Y. Sewing Machine Co., New York, Plattsburgh, N.Y.	before 1880	1882
	Later Demorest Mfg. Co.	1883	1888
Noble	Noble Sewing Machine Co., Erie, Pa.	before 1881	ca. 1887
Noble	The Noble Sewing Machine & Manu. Co. Weeping Water, Nebraska	1891	—
Novelty	C. A. French, Boston, Mass.	1869	—
Old Dominion	Old Dominion Sewing Machine Co., Richmond, Va.	ca. 1858	1860
Ormond	Ormond Manu. Co., Baltimore, Md.	1879	—
Pardox	——	ca. 1865	—
Parham	Parham Sewing Machine Co., Philadelphia, Pa.	ca. 1869	ca. 1871
Parker	Charles Parker Co., Meriden, Conn. Later Parker Sewing Machine Co.	before 1860	after 1867
Pearl	—— Bennett	ca. 1859	—
Philadelphia	Philadelphia Sewing Machine Co., Philadelphia, Pa.	ca. 1872	ca. 1881
Post Combination	Post Combination Sewing Machine Co., Washington, D.C. manufactured in Chicopee, Mass.	before 1885	1886–1888
Pratt's Patent (fig. 114) Later Ladies Companion	——	1857	ca. 1858
Premium	Foley & Williams, see Goodrich	1885	ca. 1888
Queen	Dauntless Mfg. Co., Norwalk, Ohio	1881	before 1888
Queen City	Queen City Sewing Machine Co., Cincinnati, Ohio	1882	1883
Later Cincinnati			
Quaker City (fig. 116)	Quaker City Sewing Machine Co., Philadelphia, Pa.	1859	ca. 1861
Remington Empire Later Remington	Remington Empire Sewing Machine Co.	1870	1872
	E. Remington & Sons, Philadelphia, Pa.	1873	1875
	E. Remington & Sons, Illion, N.Y.	1875	1888
	Remington Sewing Machine Agency, Illion, N.Y.	1888	ca. 1894
Robertson (dolphin & cherub) (figs. 40, 41)	T. W. Robertson, New York, N.Y.	1855	after 1860
Robinson	F. R. Robinson, Boston, Mass.	1853	ca. 1855

Sewing Machine	Manufacturer or Company	First Made or Earliest Record	Discontinued or Last Record
Robinson's patent sewing machine with Roper's improvement (fig. 117)	Howard & Davis, Boston, Mass.	1855	
Later Robinson and Roper (fig. 118)	Howard & Davis, Boston, Mass.	1856	before 1860
Rotary Shuttle	Rotary Shuttle Sewing Machine Co., Foxboro, Mass.	1881	1884
	The Foxboro Manufacturing Co., Foxboro, Mass.	1885	ca. 1887
Royal St. John (formerly St. John)	Royal Sewing Machine Co., Springfield, Ohio	ca. 1883	1891
	Rockford, Illinois	1891	1894
Ruddick	——	ca. 1860	—
Secor	Secor Machine Co., Bridgeport, Conn.	1870	1876
Sewing Shears (Hendrick's patent (fig. 43)	Nettleton & Raymond, Bristol, Conn.	ca. 1859	—
Sewing Shears	American Hand Sewing Machine Co., Bridgeport, Conn.	ca. 1884	ca. 1900
Shaw & Clark Running Stitch Machine (fig. 53) Chainstitch Machine (fig. 119)	Shaw & Clark Co., Biddeford, Me.	ca. 1857	1866
Chainstitch Machine (fig. 120)	Shaw & Clark Co., Chicopee Falls, Mass.	1867	1868
	Chicopee Sewing Machine Co., Chicopee Falls, Mass.	1868	ca. 1869
Sigwalt (see Diamond)	Sigwalt Sewing Machine Co., Chicago, Ill.	ca. 1879	1880
Singer (figs. 28–30, 32, 33, 121, 122	I. M. Singer & Co. (later Singer Mfg. Co.). Moved from Boston to New York to Elizabethport, N.J. (factory). (Some machines manufactured in the U.S. today)	1851	*
Splendid	Parsons Manufacturing Co., Chicago, Ill.	1888	—
Springfield	Springfield Sewing Machine Co., Springfield, Mass.	1880	1882
Standard (chainstitch) (fig. 123)	——	1870	—
Standard (shuttle)	Standard Shuttle Sewing Machine Co., New York, N.Y.	1874	ca. 1881
Standard	Standard Sewing Machine Co., Cleveland, Ohio (acquired by Singer Co.)	1884	1931

* Still in existence.

Sewing Machine	Manufacturer or Company	First Made or Earliest Record	Discontinued or Last Record
Stewart	Henry Stewart & Co., New York, N.Y.	1874	1883
Later New Stewart	Stewart Mfg. Co.	1880	ca. 1883
St. John (later Royal St. John)	St. John Sewing Machine Co., Springfield, Ohio	1870	ca. 1883
Taggart & Farr (figs. 124, 125)	Taggart & Farr, Philadelphia, Pa.	1858	—
Thompson	C. F. Thompson Co.	1871	1871
	T. C. Thompson, Ithaca, N.Y.	ca. 1854	
Tibbles	Tibbles Mfg. Co., Chicago, Ill.	in 1888	—
Triune	Triune Sewing Machine Co., Philadelphia, Pa.	——	out by 1888
Union	Johnson, Clark & Co., Orange, Mass.	1876	—
Union	Union Mfg. Co., Toledo, Ohio	in 1888	—
United States Family	United States Sewing Machine Co., New York, N.Y.	by 1876	ca. 1880
Victor	Finkle & Lyon Mfg. Co.	1867	ca. 1872
	Victor Sewing Machine Co., Middletown, Conn.	ca. 1872	by 1888
Wardwell	Wardwell Mfg. Co., Saint Louis, Mo.	ca. 1876	by 1888
Watson (fig. 126)	Jones & Lee	1850	ca. 1853
	Watson & Wooster, Bristol, Conn.	ca. 1853	ca. 1860
Waterbury	Waterbury Co., Waterbury, Conn.	1853	ca. 1860
Weed	T. E. Weed & Co. (became Whitney & Lyons)	1854	—
Weed	Weed Sewing Machine Co. (reorganized from Whitney & Lyons), Hartford, Conn.	1865	—
Family Favorite		1867	—
Manu. Favorite		1868	—
General Favorite		1872	—
Hartford		1881	ca. 1900
Wesson	Farmer & Gardner Manufacturing Co.	1879	1880
	D. B. Wesson Sewing Machine Co., Springfield, Mass.	1880	—
West & Willson (fig. 127)	West & Willson Co., Elyria, Ohio	1858	—
A. B. Wilson (fig. 23)	E. E. Lee & Co., New York, N.Y.	1850	1852
A. B. Wilson's patent seaming lathe Later Wheeler & Wilson (figs. 26, 27, 128, 129)	Wheeler, Wilson & Co., Watertown, Conn., then	late 1851	1853
	Wheeler & Wilson Mfg. Co., Watertown, Conn.	1853	1856
	Wheeler & Wilson Mfg. Co., Bridgeport, Conn.	1856	1905

Sewing Machine	Manufacturer or Company	First Made or Earliest Record	Discontinued or Last Record
	Singer Co., Bridgeport, Conn.	1905	1907
	Wheeler & Wilson Mfg. Co., Bridgeport, Conn.	1856	1905
	Singer Co., Bridgeport, Conn.	1905	1907
White (fig. 130)	White Sewing Machine Co., Cleveland, Ohio (machine now manufactured in Japan)	1876	*
Whitehill	Whitehill Mfg. Co., Milwaukee, Wis.	ca. 1875	by 1888
Whitney	Whitney Sewing Machine Co., Paterson, N.J.	ca. 1872	ca. 1880
Whitney & Lyons	Whitney & Lyons (a machine based on the 1854 patent of T. E. Weed)	ca. 1859	ca. 1865
Wickersham	Butterfield & Stevens Mfg. Co., Boston, Mass.	1853	—
Willcox & Gibbs (figs. 39, 131)	Willcox & Gibbs Sewing Machine Co., New York, N.Y.	1857	1973
Williams & Orvis	Williams & Orvis Sewing Machine Co., Boston, Mass.	ca. 1859	after 1860
Wilson (fig. 89) (see Buckeye)	Wilson [W. G.] Sewing Machine Co., also Star Shuttle, Cleveland, Ohio	ca. 1867	after 1885
Wilson	Wilson Sewing Machine Co.		
	Chicago	1879	1882
	Wallingford, Conn.	1882	1886
Windsor (one thread)	Vermont Arms Co., Windsor, Vt.	1856	1858
Windsor	Lamson, Goodnow & Yale, Windsor, Vt.	1859	1861
Name Unknown	John W. Beane	1853	—
"	Henry Brind	1860	—
"	Garfield Sewing Machine Co.	1881	—
"	Geneva Sewing Machine Co.	1880	—
"	Gove & Howard	1855	—
"	Charles W. Howland, Wilmington, Del.	ca.1860	—
"	Miles Greenwood & Co., Cincinnati, Ohio	ca.1861	—
"	Hood, Batelle & Co.	1854	1854
"	Wells & Haynes	1854	1854
"	Wilson H. Smith, Birmingham, Conn.	ca.1860	—
"	Continental Manufacturing Co., New York, N.Y.	—	—

* Still in existence.

In the *Sewing Machine News*, vol. 3, no. 5, p. 12 (1881), there were listed a number of then "defunct" machines and companies. Among these are many well-known names and little-known names for which at least one additional reference can be found. There are some, however, for which this is the only reference to date. These are: Blanchard, Babcock, Banner, Cottage, Cole, Duplex, Economist, Erie, Gutman, Hill, Hancock & Bennett, Jenks, Lockmar, La Favorite, Learned, Leggett, McCoy, McCardy, Medallion, McArthur & Co., Monopoly, Moreau, Mack, Niagra, New Cannaan, Orphean, Pride-of-the-West, Seamen & Guiness, Surprise, Stackpole, Shanks, Stanford, Troy, Utica, Weaver, Wagner, and Williams. Some of these names may have been a "special" name given to machines manufactured by one of the known companies, but at least a few are names of machines manufactured for a very short time prior to 1881 about which we would like to know more.

Figure 68.—AMERICAN BUTTONHOLE, Overseaming & Sewing Machine of about 1870. Using serial numbers, these machines can be dated approximately as follows: 1—7792, 1869; 7793—22366, 1870; 22367—42488, 1871; 42489—61419, 1872; 61420—75602, 1873; 75603—89132, 1874; 89133—103539, 1875; and 103540—121477, 1876. Figures are not available for the years from 1877 to 1886. (Smithsonian photo 46953–E.)

Figure 69.—(NEW) AMERICAN SEWING MACHINE of about 1874. Illustration is from a contemporary advertising brochure. (Smithsonian photo 33507.)

Figure 70.—AMERICAN MAGNETIC SEWING MACHINE, 1854. Machines of this type were manufactured for only two years under the patent of Thomas C. Thompson, March 29, 1853, and later under the patents of Samuel J. Parker, April 11, 1854, and Simon Coon, May 9, 1854. On September 30, 1853, Elias Howe listed receipts of $1000 from the American Magnetic Sewing Machine Co. for patent infringement. The machines manufactured after that date carry the Howe name and 1846 patent date to show proper licensing. Judging by Howe's usual license fee of $25 per machine, about 40 machines were manufactured prior to September 1853. The company was reported to have made about 600 machines in 1854 before it went out of business. The only American Magnetic machine known to be in eixstence is in the collection of the Northern Indiana Historical Society at South Bend, Indiana. (*Photo courtesy of the Northern Indiana Historical Society.*)

Figure 71.—ATLANTIC SEWING MACHINE, 1869. This machine is typical of the many varieties manufactured for a very short time in the 1860s and 1870s. It is about the size of the average hand-turned variety, 8 by 10 inches, but lighter in weight. The frame design was the patent of L. Porter, May 11, 1869, and the mechanism was patented by Alonzo Porter, February 8, 1870. The latter patent model bears the painted legend "Atlantic" and is stamped "Aprl 1, 69," indicating that it was probably already in commercial production. This date possibly may refer also to L. Porter's design patent, since actual date of issue was usually later than date of application. (Smithsonian photo 48329–A.)

Figure 72.—A. Bartholf sewing machine, 1853. Abraham Bartholf of New York began manufacturing Blodgett & Lerow machines (see fig. 20) about 1850; the style and mechanics of these machines, however, were primarily those of the Blodgett & Lerow patent as manufactured by O. C. Phelps and Goddard, Rice & Co. For this reason they are considered Blodgett & Lerow—not Bartholf—machines.

The true Bartholf machine evolved when the manufacturer substituted Howe's reciprocating shuttle for the rotary shuttle of the Blodgett & Lerow machine, continuing to manufacture the machine in his own adapted style. Bartholf manufactured reciprocating-shuttle machines as early as 1853, and his was one of the first companies licensed by Howe.

All Bartholf machines licensed under Howe's patent carry the Howe name and patent date. They are sometimes mistakenly referred to as Howe machines, but they are no more Howe machines than those manufactured by Wheeler & Wilson, Singer, or many others.

On April 6, 1858, Bartholf was granted a patent for an improvement of the shuttle carrier. He continued to manufacture sewing machines under the name "Bartholf Sewing Machine Co." until about 1865.

Using serial numbers, Bartholf machines can be dated approximately as follows:

Serial Number	Year	Serial Number	Year
1—20	1850	291—321	1855
21—50	1851	322—356	1856
51—100	1852	357—387	1857
101—235	1853	388—590	1858
236—290	1854	591—1337	1859

No record of the number of machines produced by Bartholf after 1859 is available.

The Bartholf machine illustrated bears the serial number 128 and the inscription "A. Bartholf Manfr, NY—Patented Sept. 1846 E. Howe, Jr." This machine is in the collection of the Baltimore County Historical Society. Note the close similarity between it and the 1850 Blodgett & Lerow machine manufactured by Bartholf. (*Photo courtesy of the Baltimore County Historical Society.*)

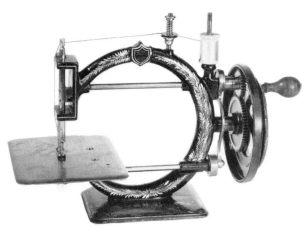

Figure 73.—Bartlett sewing machine, 1867. The Bartlett machine was first manufactured in 1866 under the January 31, 1865, and October 10, 1865, patents of Joseph W. Bartlett. The machines were made by Goodspeed & Wyman for the Bartlett Co. and were so marked. The inventor received another patent on April 7, 1868, and later machines carry this third date also. Although the first few hundred machines did not bear the dates of patents held by the "Combination," before the end of the first year of production Bartlett was paying royalties. He continued to manufacture sewing machines until the early seventies when he converted to the manufacturing of street lamps.

Using serial numbers, Bartlett's machines can be dated approximately as follows: 1—1000, 1866; 1001—3126, 1867; 3127—?, 1868. There is no record of serial numbers for the succeeding years. (Smithsonian photo 45524–G.)

Figure 74.—Bartram & Fanton sewing machine, 1867. These machines were first manufactured in 1867 under the patents of W. B. Bartram, notably his patent of January 1, 1867. Three machines were exhibited at The Eleventh Exhibition of the Massachusetts Charitable Mechanics Association in 1869 where they were awarded a bronze medal. They were compared favorably to the Willcox & Gibbs machine (see fig. 39), which they resembled. Bartram received additional patents in the early seventies and also manufactured lock-stitch machines.

Using serial numbers, machines may be approximately dated as follows: 1—2958, 1867; 2959—3958, 1868; 3959—4958, 1869; 4959—5958, 1870; 5959—6962, 1871; 6963—7961, 1872; 7962—8961, 1873; and 8962—9211, 1874. (Smithsonian photo P63198.)

Figure 75.—BECKWITH SEWING MACHINE, 1871. Among the inventors whose patent claims were "to produce a cheap and effective sewing machine" was William G. Beckwith. His machine was first manufactured by Barlow & Son, and it realized considerable success in the few years of its production. The earliest model was operated like a pair of scissors or with a cord and ring as illustrated. Beckwith later added a hand crank. The machine was purchased in Crewe, Cheshire, England; it is stamped "Pat. April 18, 71 by Wm. G. Beckwith, Foreign Pats. Secured, Barlow & Son Manuf. N.Y., [serial number] 706." By 1874 the machines were marked "Beckwith S.M. Co." and two 1872 patent dates were added.

Using serial numbers, machines may be dated approximately as follows: 1—3500, 1871; 3501—7500, 1872; 7501—12500, 8173; 12501—18000, 1874; 18001—23000, 1875; 23001—?, 1876. (Smithsonian photo 46953–C.)

Figure 76.—BOUDOIR SEWING MACHINE, 1858. This machine, a single-thread, chainstitch model was based on the patents of Daniel Harris, dated June 9, 1857, June 16, 1857, and October 5, 1858. Manufactured primarily by Bennett in Chicago in 1859, it also may have been produced in the East, although no manufacturer's name can be found.

In 1860, the Boudoir, also called Harris's Patent sewing machine, was exhibited at the Massachusetts Charitable Mechanics Association Exhibition where it won a silver medal for "its combination of parts, its beauty and simplicity, together with its ease of operation." At this time the machine was described as making a "double lock stitch" (another name for the double chainstitch). It was also described as having been before the public for some time and combining "the improvements of others for which the parties pay license." The machine head was positioned on the stand similarly to that of the West & Willson (fig. 127) and stitched from left to right.

It is not known exactly how many of these machines were made or how long they were in vogue. Manufacture, although probably ceasing in the 1860s, is known to have been discontinued before 1881, when a list of obsolete sewing machines was published in *The Sewing Machine News*. (Smithsonian photo P63199.)

Figure 77.—(New) Buckeye sewing machine of about 1875. The Buckeye machine was one of several manufactured by W. G. Wilson of Cleveland, Ohio. It was licensed under Johnson's extended patent of April 18, 1867. Although it was small and hand turned, it used two threads and a shuttle to form a lockstitch. The machine was sufficiently popular for Wilson to introduce an improved model in the early 1870s, which he called the New Buckeye. W. G. Wilson continued to manufacture sewing machines until about the mid-eighties, although the Buckeye machines were discontinued in the seventies. (Smithsonian photo 45524–A.)

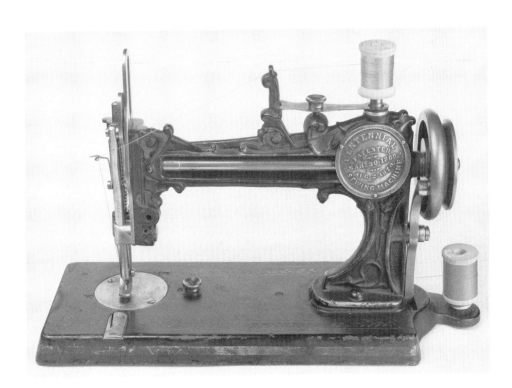

Figure 78.—Centennial sewing machine, 1876. The Centennial machine was basically a McLean and Hooper sewing machine which was renamed to take advantage of the coming Centennial celebration. It was based on the patents of J. N. McLean, March 30, 1869, and August 2, 1870, and made a two-thread chainstitch. Only about five hundred Centennial machines were manufactured in 1873, but by 1876 over three thousand had been constructed. The machines were advertised on white circulars which were printed in red and blue, and engraved with two women sewing, one by hand, labeled "Sewing in 1776," and one at a Centennial sewing machine, labeled "Sewing in 1876." There is no record that the machines were made after 1876. (Smithsonian photo 48216–T.)

Figure 79.—CLARK's Revolving-Looper double-thread sewing machine, 1860. This machine was manufactured by Lamson, Goodnow, & Yale of Windsor, Vermont. It was an attempt to improve on the combined ideas of the Grover and Baker machine, the Nettleton & Raymond machine, and the earlier single-thread Windsor machine. The improvements were made and patented by Edwin Clark on December 6, 1859. Widely advertised, the machines sold for $35 with a foot-power table. They could also be operated by hand. Over three thousand were manufactured and sold, and preparations were being made to continue manufacture of the earlier single-thread Windsor, originally made by the company's predecessor, Vermont Arms Co., when the Civil War broke out. A flood of arms orders arrived, and the sewing-machine manufacture was discontinued early in the summer of 1861. The sewing-machine equipment and business was sold to Grout & White of Massachusetts. (Smithsonian photo 48216.)

Figure 80.—DU LANEY SEWING MACHINE of about 1872. Most of the small, simple, chainstitch sewing machines of this period were constructed so that they could either be turned by hand or set into a treadle-powered table. Du Laney's Little Monitor, manufactured for only a few years, was based on the patents of G.L. Du Laney, July 3, 1866, and May 2, 1871. It was a two-thread, chainstitch machine powered only by a foot treadle. By simple adjustment, the machine could also make the cablestitch and the lockstitch. (Smithsonian photo 48221–C.)

Figure 81.—EUREKA SHUTTLE SEWING MACHINE, 1859. An example of the many short-lived types of which no written record can be found, this particular machine was used as a patent model for certain minor improvements in 1859. It has the name "Eureka" painted on the top and the following inscription incised on the baster plate: "Eureka Shuttle Sewing Machine, 489 Broadway, New York." Although it is a shuttle machine, this one carries no patent dates. A second style shuttle machine was introduced in late 1859 and continued to be manufactured in 1860. Charges of infringement must have forced the company to pay royalty costs. A machine of 1860 has the serial number 1925 and many of the patent dates held by the Combination. The forced payment for the use of the patents may have caused the failure of the company. After moving from their first address to 493 Broadway, the company disappeared by 1861. (Smithsonian photo 48328–C.)

Figure 82.—M. FINKLE SEWING MACHINE, 1857. The M. Finkle machines were manufactured in 1856 and 1857. Sometime before or about 1859, the inventor, Milton Finkle, formed a partnership and the machines were subsequently called M. Finkle & Lyon and later simply Finkle & Lyon. In 1859 the machine was awarded a silver medal by the American Institute for producing superior manufacturing and family lockstitch sewing machines. It also won a silver medal in Boston in 1860 at the Massachusetts Charitable Mechanics Association Exhibition. Although the name of the machine was changed to Victor in 1867, the company name remained Finkle & Lyon until about 1872 when it was changed to Victor also. Victor machines were manufactured until about 1890.

Machines can be dated by their serial number approximately as follows:

Serial Number	Year	Serial Number	Year
1—200	1856	13001—15490	1867
201—450	1857	15491—17490	1868
451—700	1858	17491—18830	1869
701—950	1859	18831—21250	1870
951—1500	1860	21251—28890	1871
1501—3000	1861	28891—40790	1872
3001—5000	1862	40791—48240	1873
5001—7000	1863	48241—53530	1874
7001—9000	1864	53531—59635	1875
9001—11000	1865	59636—65385	1876
11001—13000	1866		

No estimates are available for the years 1877 to 1890. (Smithsonian photo 48216–A.)

Figure 83.—FLORENCE SEWING MACHINE. The Florence machine was based on the patents of Leander W. Langdon, whose first patent was obtained in 1855. Langdon sewing machines were manufactured by the inventor for a few years. It was his patent of March 20, 1860, that was the immediate forerunner of the Florence machine, whose name was derived from the city of manufacture, Florence, Massachusetts. The Howe royalty records of 1860 listed the Florence Sewing Machine Co. as one that took out a license that year. Langdon's patent of July 14, 1863, was incorporated into the machines manufactured after that date; however, the date is always incorrectly stamped "July 18, 1863." In 1865, the machine won a silver medal at the Tenth Exhibition of the Massachusetts Charitable Mechanics Association.

Over 100,000 Florence machines were manufactured by 1870. About 1880 the company changed the name of the machine to Crown. Improvements led to the name New Crown by 1885. About this time the right to use the name Florence for a sewing machine was purchased by a midwestern firm for an entirely different machine. In 1885 the Florence company began to manufacture lamp stoves and heating stoves and shortly thereafter they discontinued the manufacture of sewing machines.

Using the serial numbers, Florence machines can be dated approximately as follows:

Serial Number	Year	Serial Number	Year
1—500	1860	82535—96195	1869
501—2000	1861	96196—113855	1870
2001—8000	1862	113856—129802	1871
8001—20000	1863	129803—145592	1872
20001—35000	1864	145593—154555	1873
35001—50000	1865	154556—160072	1874
50001—60000	1866	160073—164964	1875
60001—70534	1867	164965—167942	1876
70535—82534	1868		

No record of the number of machines produced each year between 1877 and 1885 is available.

The machine shown here, serial number 49131, was manufactured in 1865. It is stamped with the following patent dates: "Oct. 30, 1855, Mar. 20, 1860, Jan. 22, 1861, and July 18, 1863" and the Wilson patent date "Nov. 12, 1850." The machines from 1860–1863 are marked with the early Langdon patents, excluding the 1863 one, and they have the additional patent dates of Howe and others: "Sept. 10, 1846, Nov. 12, 1850, Aug. 12, 1851, May 30, 1854, Dec. 19, 1854, Nov. 4, 1856." (Smithsonian photo 45572–A.)

Figure 84.—GLOBE SEWING MACHINE. J. G. Folsom received two design patents in 1864, one on March 1 for a spool holder and one on May 17 for the basic style of the machine. Also in the same year, he was awarded a mechanical patent for an adjustment in the lower looper that would accommodate a change in needle size. Using these patents, he manufactured a single-thread, chainstitch machine, the Globe. Folsom also exhibited his machines at the Tenth Exhibition of the Massachusetts Charitable Mechanics Association in 1865. The Globe attracted particular attention and was awarded a silver medal.

In 1866 Folsom devised a new treadle attachment for hand-operated machines; the invention was featured in *Scientific American*, volume 14, number 17, with a Globe machine. Folsom again exhibited at the Massachusetts Mechanics exhibition in 1869. In addition to an improved single-thread Globe, he also showed a double-thread, elastic-stitch (double chainstitch) machine for which he received a silver medal.

Folsom machines were manufactured until 1872; 280 machines were manufactured in 1871.

The Globe sewing machine illustrated is stamped "J. G. Folsom, Maker, Winchendon, Mass. Patented April 28, 1863 [Ketchum's patent], Mar. 1, 1864. May 17, 1864." The machine was manufactured before November 1864 or it would include the patent for the lower loop adjustment. (Smithsonian photo 48216–H.)

NOTE: At least five sewing machines, those in figures 84 through 89, are similar enough in appearance to cause some confusion, because their basic design stems from a short pillar.

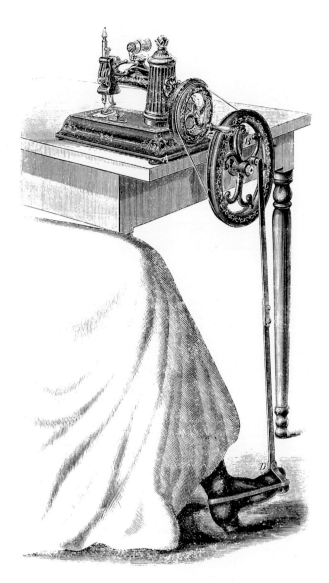

Figure 86.—EMPIRE SEWING MACHINE, late 1860s. Although an Empire Sewing Machine Co. existed in New York in the 1860s (the predecessor of the Remington-Empire Co.), it is not known whether this machine was manufactured by that same company, which was primarily concerned with producing shuttle machines. This chainstitch machine is marked "Empire Co., Patented April 23, 1863," the date referring again to Ketchum's patent. It is very similar to Folsom's Globe, except that it has claw feet rather than a closed base; the painted designs on the base of both are almost identical to those on the Monitor. Its spool holder, mounted in reverse, is a crude imitation of the Folsom patent. The Empire machines were probably manufactured about the same time as the Wilson machine. (*Photo courtesy of The Henry Ford Museum and Greenfield Village, Dearborn, Michigan.*)

Figure 85.—GLOBE SEWING MACHINE with treadle attachment as illustrated in *Scientific American*, April 21, 1866. (Smithsonian photo 48221–A.)

Figure 87.—ATWATER SEWING MACHINE, 1858. Atwater machines, based on the patent of B. Atwater, issued May 5, 1857, were manufactured from 1857 to about 1860. The machine illustrated, which is designed to be operated by a hand-turned wheel, has an upper forked dog feed, and its horizontally supported spool is directly over the stitching area. Like the others, it has a striated pillar and claw feet. The manufacturer is unknown. (Smithsonian photo P63200.)

Figure 88.—MONITOR SEWING MACHINE, 1860–1864. The Monitor machines of this style were not marked by their manufacturers, Shaw & Clark of Biddeford, Maine. Later the company was forced by the "Combination" to pay a royalty, so it changed the style and began marking its machines with the company name and patent dates (see fig. 119 for copy of seal). The Monitor, which employed the conventional vertical spindle to hold the spool of thread, had a top feed in the form of a walking presser. Its striated pillar was similar to that of the Atwater machine, and both featured the same claw feet and urn-like top. Unlike the Atwater, however, the Monitor had a double drive from the hand-turned wheel, which was grooved for operation with belt and treadle. (Smithsonian photo 33458.)

Figure 89.—WILSON SEWING MACHINE, late 1860s to early 1870s. In addition to the Buckeye (see fig. 77), W. G. Wilson manufactured several other styles of sewing machines. This one, a combination of the varying styles of the earlier pillar machine has even duplicated the general style of the spool holder patented by Folsom. The pillar is not striated, but the machine does repeat the claw feet of the Atwater and Monitor machines. Wilson machines are usually marked "Wilson Sewing Mach. Manuf'g Co. Cleveland, Ohio, Ketchum's Patent April 28, 1863." The latter name and/or patent date are found on many of the machines of this general construction. The patent is that issued to Stephen C. Ketchum for his method of converting rotary motion into reciprocal motion. (*Photo courtesy of The Henry Ford Museum and Greenfield Village, Dearborn, Michigan.*)

Figure 90.—Grant Brothers sewing machine, 1867. This machine was one of several styles that utilized Raymond's 1861 patented chainstitch method. This machine, however, used an under feed rather than a top feed.

Neither a name nor a date appears on the machine. In the June 25, 1907, issue of the *Sewing Machine Times* it was called the Common Sense machine, but detailed research has turned up no evidence to substantiate this name. However, a dated brochure advertising the Grant Brothers machine and showing a model identical to that illustrated in the *Sewing Machine Times* has been found. The brochure states that the machine made an elastic lockstitch; this was not a true lockstitch, however, but was in fact a simple chainstitch.

Grant Brothers sold their machine, which had silver-plated mountings, for $18; the price included hemmer, Barnum's self-sewer, oilcan, screwdriver, clamp, gauge, and four silver needles. An additional charge of $12 was made for a table and treadle. Compared to other chainstitch machines the price was high, and the company was short-lived. (Smithsonian photo 60794–E.)

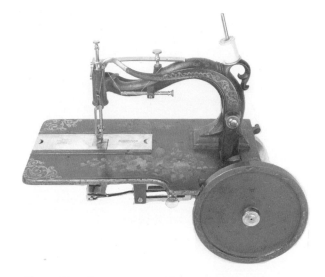

Figure 91.—Greenman and True sewing machine. This lockstitch machine based on S. H. Roper's patent of 1857 was manufactured at Norwich, Connecticut, from 1859 to 1861 by Cyrus B. True, the inventor, and Jared F. Greenman, True's financial partner. Licensed by the "Combination" and carrying the Howe patent date, the machine had obvious merit: it was strong, well made—a good family machine. Exhibited at the Ninth Exhibition of the Massachusetts Charitable Mechanics Association in September 1860, it received a bronze medal. (At this time the company was listed as Morse and True—the inventor had obviously taken on a second financial backer.) Unfortunately, the best market for the machine lay in the South, and the outbreak of the Civil War made collections impossible. This greatly retarded business and finally drove the firm into bankruptcy. In all, it is doubtful that more than one thousand machines were produced in the three years of manufacture.

The machine illustrated is marked "Greenman and True" and bears the serial number 402; it was probably manufactured early in 1860. (Smithsonian photo 48216–N.)

Figure 92.—Grover and Baker sewing machine. The Grover and Baker machine was one of the more popular machines from the 1850s until the early 1870s. The company produced iron-frame machines, fine cabinet models, and portables (figs. 35 and 36). Their machines may be dated by serial number approximately as follows:

Serial Number	Year	Serial Number	Year	Serial Number	Year
1—500	1851	5039—7000	1856	44870—63705	1861
501—1000	1852	7001—10681	1857	63706—82641	1862
1001—1658	1853	10682—15752	1858	82642—101477	1863
1659—3893	1854	15753—26033	1859	101478—120313	1864
3894—5038	1855	26034—44869	1860	120314—139148	1865
				139149—157886	1866
				157887—190886	1867
				190887—225886	1868
				225887—261004	1869
				261005—338407	1870
				338408—389246	1871
				389247—441257	1872
				441258—477437	1873
				477438—497438	1874
				497439—512439	1875

(Smithsonian photo 45513–B, an engraving of a Grover and Baker sewing machine from an advertising brochure of about 1870.)

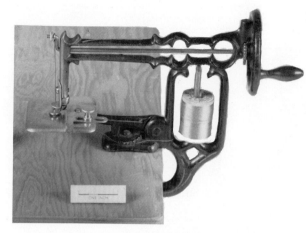

Figure 93.—Hancock sewing machine, 1867. One of the many inventors who turned his talents to inventing and producing a mechanically simple and cheaper machine was Henry J. Hancock. His 1867 machine is only about six inches wide; it uses a tambour-type needle, pulling a loop of thread from below the stitching surface. (Smithsonian photo P63197.)

Figure 94.—Hancock sewing machine, 1868. Hancock in 1868 received both a design patent and a mechanical patent now using the eye-pointed needle and a hook to form the chainstitch. The design was an open framework circle with a mirror mounted in front of the table clamp. The purpose of the designated "looking glass" was decorative only. The Hancock machines were only manufactured for a few years. They measure 10½ inches in width, slightly larger than the earlier machine. (Smithsonian photo 48328–M.)

Figure 95.—[A.C.] Herron's patent sewing machine, 1858. The general style of this machine and the last Watson machines that sold for ten dollars are very similar. The Herron machines carried the patentee's improvement in the method of making the chainstitch and bore a small heart-shaped brass plate marked "A. C. Herron's Patent Aug. 4, 1857." The machines were provided with a hand crank, but were also meant to be operated by a belt and treadle. This machine head measures 14 inches wide. The manufacturer of this machine is not known, but it may have been produced by Watson & Wooster also. No records of the extent of manufacture of the Herron machines have been found. (Smithsonian photo 48329–J.)

Figure 96.—A. B. Howe sewing machine of about 1860. (Smithsonian photo 45525–C.)

Figures 96, 97, and 98.—The Howe machines. It is difficult for many to believe that the stamped legend "Elias Howe patent, Sept. 10, 1846" does not certify that a machine is an original Howe. Although Elias Howe was granted a patent for the lockstitch machine in 1846, he did not establish a sewing-machine factory for about twenty years. Early in the 1850s and later through the "Combination," however, he licensed others to make machines using his patent. These machines bore that patent date for which a royalty was being paid.

Among his early licensees was his elder brother Amasa who organized the Howe Sewing Machine Co. in 1854. The Amasa Howe machines were very good ones, and in 1862 Amasa won the prize medal at the London International Exhibition. This immensely increased the popularity of the machine and Elias offered to join Amasa by building a large factory at Bridgeport, Connecticut, to fill the increasing demand for more machines. The machines produced at Bridgeport, however, although imitating the Amasa Howe machines, proved inferior in quality. Amasa found that, rather than helping his business reputation, his brother's efforts were hurting him, and he severed business relations with Elias.

Because of their brief association, the 1862 prize medal awarded to A. B. Howe was sometimes credited to Elias. The latter did receive awards for his patent, but never for his manufactured machines. When the two brothers dissolved their joint venture, Elias attempted to call his new company the Howe Sewing Machine Co., but Amasa's claim that this name had been his exclusive property for many years was upheld by the courts. Elias then omitted the word "Sewing" and called his company simply the Howe Machine Co.

After Elias died in 1867, the company was run by his sons-in-law, the Stockwell brothers. To distinguish their machines from those of A. B. Howe, they marked each machine with a brass medallion picturing the head and flowing locks of Elias Howe. They also continued to advertise their machine as the "original" Howe. In about 1873, B. P. Howe, Amasa's son, sold the Howe Sewing Machine Co. to the Stockwell brothers, who continued to manufacture Howe machines until 1886.

The machines of the A. B. Howe Sewing Machine Co. may be dated by serial number approximately as follows:

Serial Number	Year	Serial Number	Year
1–60	1854	167–299	1857
61–113	1855	300–478	1858
114–166	1856	479–1399	1859

No figures are available for 1860–1870, but 20,051 machines were manufactured in 1871.

The machines of the [Elias] Howe Machine Co. are not believed to have begun with serial number 1, and no figures are available for 1865–1867. After that, the machines may be dated by serial number approximately as follows:

Serial Number	Year	Serial Number	Year
11,000–46,000	1868	446,011–536,010	1873
46,001–91,843	1869	536,011–571,010	1874
91,844–167,000	1870	571,011–596,010	1875
167,001–301,010	1871	596,011–705,304	1876
301,011–446,010	1872		

No figures are available for 1877–1886.

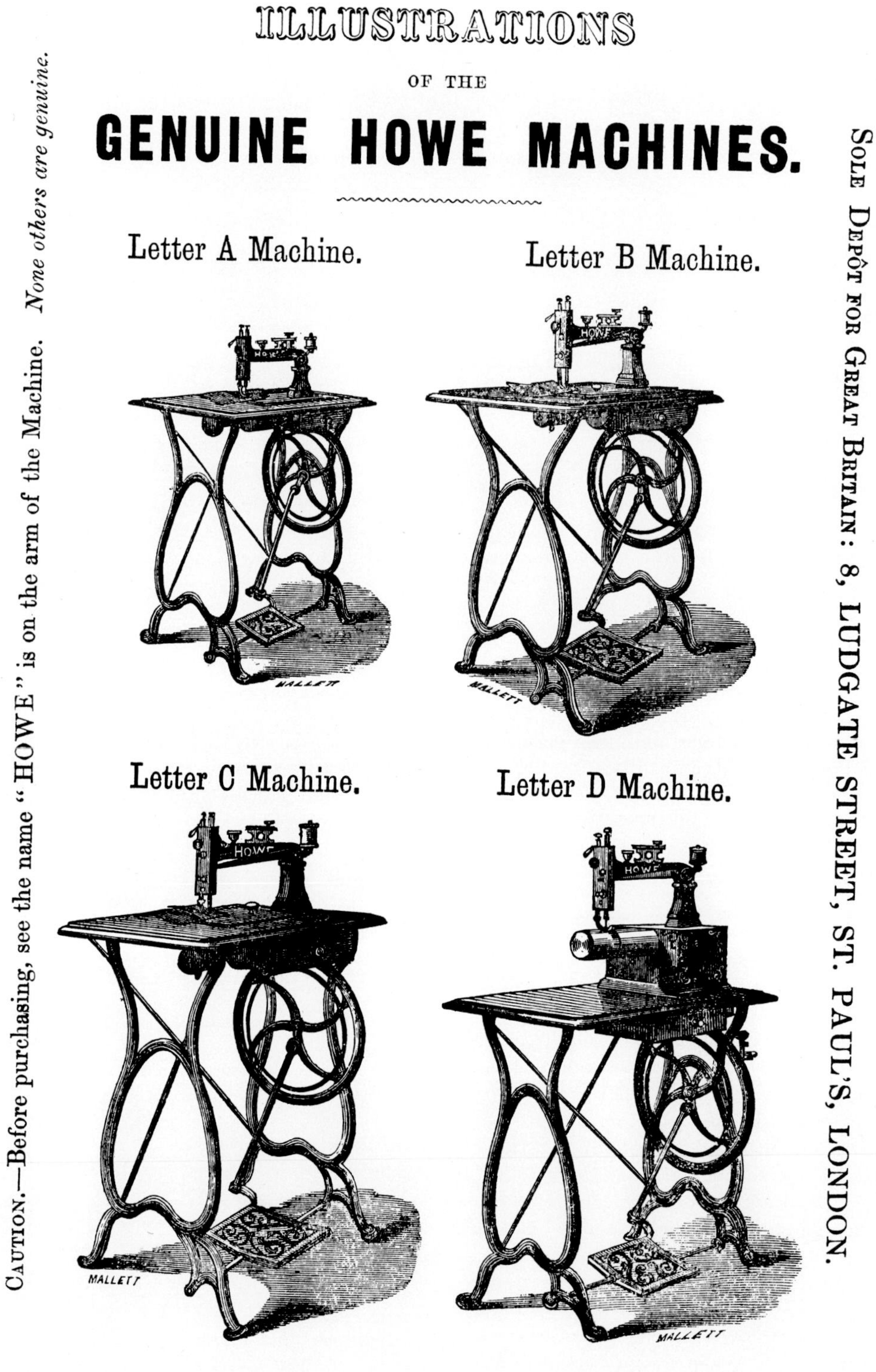

Figure 97.—ADVERTISING BROCHURE distributed by E. Howe during the brothers' brief partnership; the machines are basically A. B. Howe machines, 1863. (Smithsonian photo 49373–A.)

Figure 98.—Howe (Stockwell brothers) machine, 1870. (Smithsonian photo 45572–E.)

Figure 99.—Patent model of Christopher Hodgkins, November 2, 1852, assigned to Nehemiah Hunt. (Smithsonian photo 34551.)

Figures 99, 100, and 101.—The N. Hunt (later, in 1856, Hunt & Webster and finally in 1858 Ladd and Webster) sewing machine was based on the patents of Christopher Hodgkins, November 2, 1852, and May 9, 1854, both of which were assigned to Nehemiah Hunt. First manufactured in 1853, the machine, which closely resembled the Hodgkins' patent, won a silver medal at the exhibition of the Massachusetts Charitable Mechanics Association that same year.

In 1856 Hunt took a partner, and the company became Hunt & Webster. An interesting account of this company appeared as a feature article in *Ballou's Pictorial*, July 5, 1856, where it was reported that "the North American Shoe Company have over fifty of the latest improved machines, represented in these drawings [fig. 31], now running . . . ." The article also estimated that a 55-million dollar increase in shoe manufacturing in Massachusetts in 1855 was due to the sewing machine. In 1856 the Hunt & Webster machine again won a silver medal at the exhibition. Very late in 1858 the company became Ladd, Webster, & Co. and continued to manufacture both family and manufacturing sewing machines until the mid-1860s.

The approximate date of manufacture can be determined by serial number:

Serial Number	Year
1—100	1853
101—368	1854
369—442	1855
443—622	1856
623—1075	1857
1076—1565	1858
1566—3353	1859

No figures are available for the 1860s.

Figure 100.—RIGHT: HUNT & WEBSTER sewing machine of about 1855, serial number 414. (Smithsonian photo 48216–V.)

Figure 101.—LADD, WEBSTER & Co. sewing machine of about 1858, Boston, serial number 1497. (Smithsonian photo 46953.)

Figure 102.—IMPROVED COMMON SENSE sewing machine of about 1870. This machine is so very similar to the New England machines in its feed, threading, looping mechanism, and in its general design, that it is sometimes mistaken for the earlier New England machines (see figs. 112 and 113).

Dating from the early 1870s, the Improved Common Sense machine is about 10 inches in width, two inches larger than the New England machine. The spool holder is similar to Folsom's patented design, but is less refined. A page from an advertising brochure of the period verifies the name of the machine, but does not identify the manufacturer.

There are no patent dates or identifying names or numbers on the machine illustrated. Although the Empire Co. also produced a machine of this style, their models are marked with their name and with Ketchum's patent date, April 23, 1863. Of the several styles of machine using the Raymond looper, this type seems to account for the largest volume manufactured, as evidenced by the proportionately higher number of examples still extant. (Smithsonian photo 48328–E.)

Figure 103.—JOHNSON SEWING MACHINE, 1857. Another of the all-but-forgotten manufacturers of the 1850s was Emery, Houghton & Co., who constructed the A.F. Johnson machines. Examination of existing machines indicates that they were manufactured in 1856 and 1857, and possibly a little longer. This one from 1857 bears the serial number 624, so we know that several hundred were manufactured. The head is ornately attractive, slightly reminiscent of Wheeler & Wilson models, and of standard size. (Smithsonian photo 48329–B.)

Figure 104.—"Lady" sewing machine of about 1859. The contemporary name of this machine is unknown. The unusual design of the head, or main support, is based in part on the design patent, number 216, of Isaac F. Baker, issued April 10, 1849, for a "new and useful design [,] for ornamenting furniture [,] called Cora Munro" who was a character in James Fenimore Cooper's *Last of the Mohicans*. The design shows a female figure wearing a riding dress and hat that is ornamented with a plume and a bow. Her right hand holds a riding stick and the left, her skirt. Trunks of trees and foliage complete the Baker design, which is known to have been used for girandoles of the period. A companion design was also patented by Baker, number 215, which is in the form of a man in military costume and is named "Major Heyward," for another character in *Last of the Mohicans*.

The sewing machines based on the "Cora Munro" design also use branch designs as the overhanging arms. A mother bird sits in the upper branch and descends to feed a young bird as the machine is in operation. The one illustrated was used as the machine submitted with a request for patent by George Hensel of New York City for which patent 24,737 was issued on July 12, 1859. Since Hensel's patent application was for an improvement in the feed, there was no need for the highly decorative head unless such a machine was commercially available. The patent specifications merely state that the head is "ornamented." Another sewing machine of this type was used as the patent model by Sidney Parker of Sing Sing, New York, number 24,780, issued on the same date as the Hensel patent. Parker's patent also covered an improved feeding mechanism. In the patent description, however, the inventor states that "the general form of the machine is not unlike others now in use." By this he might have meant in the design, or possibly in the basic structural form. Other than the two machines described, no other examples are known to have survived, but "Lady" or "Cora Munro" sewing machines were manufactured. (Smithsonian photo 45506–D.)

Figure 105.—LANDFEAR'S PATENT SEWING MACHINE of about 1857. Another of the many machines that, except for isolated examples, have almost completely disappeared from the records is Landfear's machine. Fortunately, this manufacturer marked his machine—where many did not—stamping it: "Landfear's patent – Decr 1856, No. 262, W. H. Johnson's Patent Feb. 26th 1856, Manfrd by Parkers, Snow, Brooks & Co., West Meriden, Conn." (There was a Parker sewing machine manufactured by the Charles Parker Co. of Meriden, but his machine was a double-thread chainstitch machine and was licensed by the "Combination." The Landfear machine may have been an earlier attempt by a predecessor or closely related company.)

The Landfear patent was for a shuttle machine, but it also included a mode for regulating stitch length. The name chosen for this machine may be incorrect, since the single-thread chainstitch mechanism is primarily that of W. H. Johnson, but since the Johnson patent also was used on other machines the name "Landfear" was assigned. The machine was probably another attempt to evade royalty payment to the "Combination."

The serial number 262 indicates that at least that many machines were manufactured, although this model is the only one known to be in existence. The support arm of the machine head is iron, cast as a vase of flowers and painted in natural colors. The paint on the head is original, but the table has been refinished, and the iron legs, which had rusted, have been repainted. (Smithsonian photo 48440–G.)

Figure 106.—Lathrop sewing machine of about 1873. These machines were manufactured by the Lathrop Combination Sewing Machine Co. under the patents of Lebbeus W. Lathrop of 1869, 1870, and 1873. The machine used two threads, both taken from spools; moreover, it produced not only the double chainstitch, but it was constructed to produce also a lockstitch and a combined "lock and chain stitch." The machine illustrated bears the serial number 31 and the patent dates of Grover & Baker, and Bachelder among others, in addition to the first two Lathrop patent dates. The company lasted only a few years as it is included in the 1881 list of manufacturers that had ceased to exist. (Smithsonian photo 46953–F.)

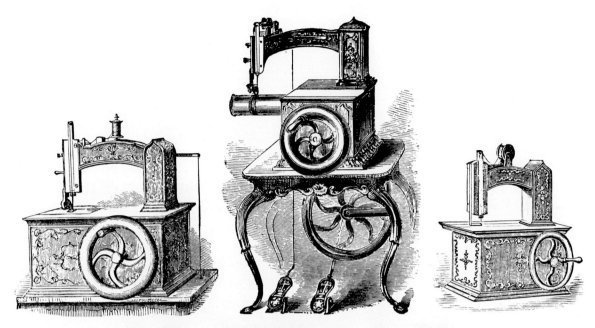

J. B. NICHOLS & CO.,
MANUFACTURERS OF
HOWES' IMPROVED PATENT

Figure 107.—Illustration from a brochure, marked in ink: "The National Portrait Gallery, 1855." Singer Archives. (Smithsonian photo 48091–E.)

Figures 107 and 108.—The Nichols and Leavitt sewing machines. One of Elias Howe's earliest licensees was J. B. Nichols. His machine, manufactured at first with George Bliss and later alone as J. B. Nichols & Co., was called Howe's Improved Patent Sewing Machine. It was, however, no more a Howe machine than any of the others produced under the Howe patent.

In July 1855 Nichols went into partnership with Rufus Leavitt, and the company name changed to Nichols, Leavitt & Co. In 1857 it was changed again to Leavitt & Co., and finally in the mid-1860s to Leavitt Sewing Machine Co. By the 1870s, it was defunct.

The Nichols-Leavitt machines can be dated by their serial numbers approximately as follows:

Serial Number	Year	Company
1—28	1853	Nichols & Bliss
29—245	1854	J. B. Nichols & Co.
246—397	1855	J. B. Nichols & Co.—Nichols, Leavitt & Co.
398—632	1856	Nichols, Leavitt & Co.
633—827	1857	Leavitt & Co.
828—902	1858	"
903—1115	1859	"
1116—1436	1860	"
1437—1757	1861	"
1758—2077	1862	"
2078—2400	1863	"
2401—2900	1864	"
2901—3900	1865	Leavitt Sewing Machine Co.
3901—4900	1866	"
4901—5951	1867	"
5952—6951	1868	"
6952—7722	1869	"

There is no record that the company was in existence after 1869.

Figure 108.—LEAVITT SEWING MACHINE of about 1868, serial number 6907. (Smithsonian photo 48328.)

Figure 109.—Lester sewing machine of about 1858. The Lester machine was first manufactured by J. H. Lester in Brooklyn, New York. His machine was based on the patents of William Johnson, John Bradshaw and others but not on the patents held by the "Combination," although he had secured a license. When the Old Dominion Company applied for a license from the "Combination," Lester learned of this, went to Richmond, and arranged to combine his business with theirs. Since the Lester machine was the better one, it was agreed to cease the manufacture of the Old Dominion machines early in 1860 and in March the company name was changed to the Lester Mfg. Co. Late in 1860, George Sloat entered the company with his Elliptic machine; the name was changed again, this time to Union Sewing Machine Co. The manufacture of both sewing machines continued until the outbreak of the Civil War the following year, which brought a conversion to arms production. The manufacture of Lester machines was never resumed.

The machine illustrated was manufactured by J. H. Lester in Brooklyn; it bears the serial number 96. The number of Lester machines manufactured from 1858 through 1861 is not known, but it was probably less than 1,000. (Smithsonian photo P63359.)

Figure 110.—Ne Plus Ultra of about 1857. Another of the interesting hand-turned chainstitch machines of the late 1850s and 1860s was patented by O.L. Reynolds. The baster plates and the handle on the wheel are missing on this machine, but an interesting shield and draped-flag pattern is painted on the base.

Another machine of this type has the following inscription stamped on the baster plate: "Ne Plus Ultra, Patent Applied For, 174, O.L. Reynolds, Patentee & Manufacturer, Dover N.H." Reynold's patent model, March 30, 1858, bears the serial number 110, indicating that the machine illustrated here—which bears the serial number 26—was manufactured before the patent was obtained. (Smithsonian photo 48216–F.)

Figure 111.—Nettleton & Raymond sewing machine. One of the most ornate of the early, small, hand-turned sewing machines was patented and manufactured by Willford H. Nettleton and Charles Raymond whose first patent was received on April 14, 1857. The patent model, believed to be a commercial machine, is beautifully silver-plated. Whether this was a special one-of-a-kind model, or whether the inventors tried to make a commercial success of a silver-plated machine is not known. The machine made a two-thread chainstitch, taking both threads from commercial spools. By October 1857, the inventors had received their second patent. This time the machine was brass and gilt—brighter, but less expensive. At the same time, Nettleton & Raymond began manufacturing sewing-shears machines under the patent of J. E. Hendricks.

By the latter half of 1858, Nettleton & Raymond had moved from Bristol, Connecticut, to Brattleboro, Vermont. The patented improvement of the two-thread chainstitch machine received that year was in the name of "Raymond, assignor to Nettleton," although the machines of this type bear neither name nor patent date. No record of the price for which they were sold has been found, but it would be fair to estimate that it was probably about $25. This style of machine was discontinued when the manufacture of the simpler, more profitable New England model began, a machine that Raymond had initiated just before the partners left Bristol. (Smithsonian photo 45505-E.)

Figure 112.—Raymond patent model, March 9, 1858. (Smithsonian photo 32009–0.)

Figures 112 and 113.—New England sewing machines. The small, hand-turned, sewing machines some of which were called Common Sense, were manufactured by at least three companies and possibly more. The earliest ones were those made by Nettleton & Raymond based on Charles Raymond's patent of March 9, 1858, which featured a hinged presser foot acting as the top feed. On July 30, 1861, Raymond received a patent for an improved looper; this date is found on all machines later manufactured by the inventor.

In 1858 Nettleton and Raymond had moved from Bristol, Connecticut, to Brattleboro, Vermont. Also in Brattleboro at this time were Thomas H. White and Samuel Barker, who were manufacturing a small machine called the Brattleboro. White left Vermont in 1862 and went to Massachusetts. There, in partnership with William Grout, he also began to manufacture New England machines; these were basically the same as the Raymond machines. After a short time, Grout left the partnership with White and moved to Winchendon, there continuing to make New England machines for approximately one more year. In 1865, J. G. Folsom of Winchendon exhibited a New England machine at the Tenth Exhibition of the Massachusetts Charitable Mechanics Association along with his Globe machine. Whether both machines were manufactured by him or whether he might have been exhibiting one of Grout's machines is not known.

There is no record that New England machines were manufactured after 1865. There is a great similarity between these machines and the Improved Common Sense sewing machines of the 1870s. It is believed that the name "Common Sense" was given by frugal New Englanders to several of the cheaper chainstitch machines of the 1860s.

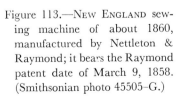

Figure 113.—New England sewing machine of about 1860, manufactured by Nettleton & Raymond; it bears the Raymond patent date of March 9, 1858. (Smithsonian photo 45505–G.)

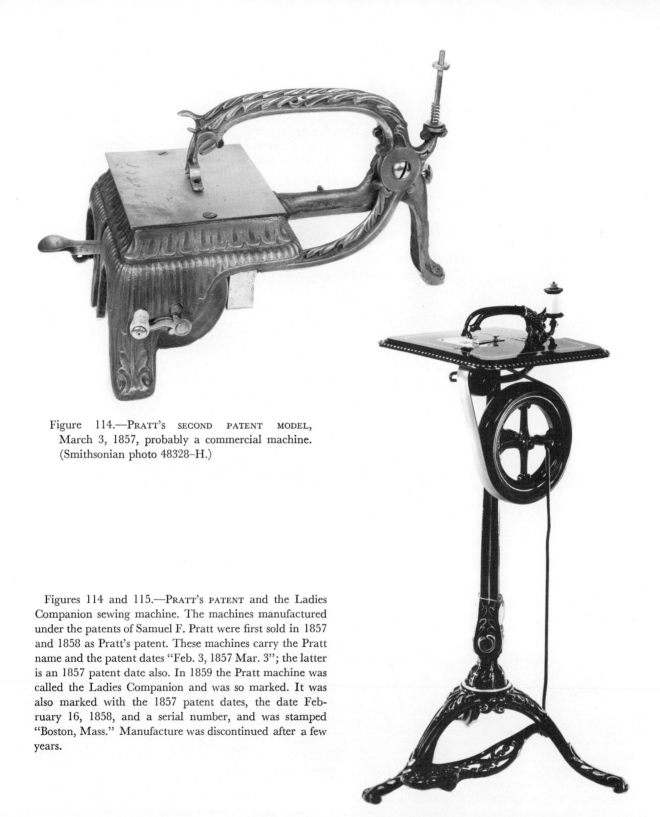

Figure 114.—Pratt's second patent model, March 3, 1857, probably a commercial machine. (Smithsonian photo 48328–H.)

Figures 114 and 115.—Pratt's patent and the Ladies Companion sewing machine. The machines manufactured under the patents of Samuel F. Pratt were first sold in 1857 and 1858 as Pratt's patent. These machines carry the Pratt name and the patent dates "Feb. 3, 1857 Mar. 3"; the latter is an 1857 patent date also. In 1859 the Pratt machine was called the Ladies Companion and was so marked. It was also marked with the 1857 patent dates, the date February 16, 1858, and a serial number, and was stamped "Boston, Mass." Manufacture was discontinued after a few years.

Figure 115.—Ladies Companion, 1859. (*Photo courtesy of The Henry Ford Museum and Greenfield Village, Dearborn, Michigan.*)

Figure 116.—Quaker City sewing machine. During the first decade of sewing-machine manufacture many types of handsome wooden cases were developed to house the mechanisms. Although such cases increased the total cost, they were greatly admired and were purchased whenever family funds permitted. The machine was based on the patents of William P. Uhlinger: a mechanical patent for a double chainstitch machine on August 17, 1858 (antedated May 8), and a patent for the casing on December 28, 1858. The machine head was lowered into the casing as the lid was brought forward and closed—an idea much ahead of its time.

This Quaker City machine, serial number 18, was purchased by Benjamin F. Meadows of Lafayette, Alabama, for $150 just prior to the Civil War. Relatively few machines of this type were manufactured, and the Quaker City Sewing Machine Co. existed for only a few years. Its apparent hope for a southern market was short-lived, and it was unable to compete either with the companies licensed under the "Combination" or with those producing less expensive machines. (Smithsonian photo 46953–A.)

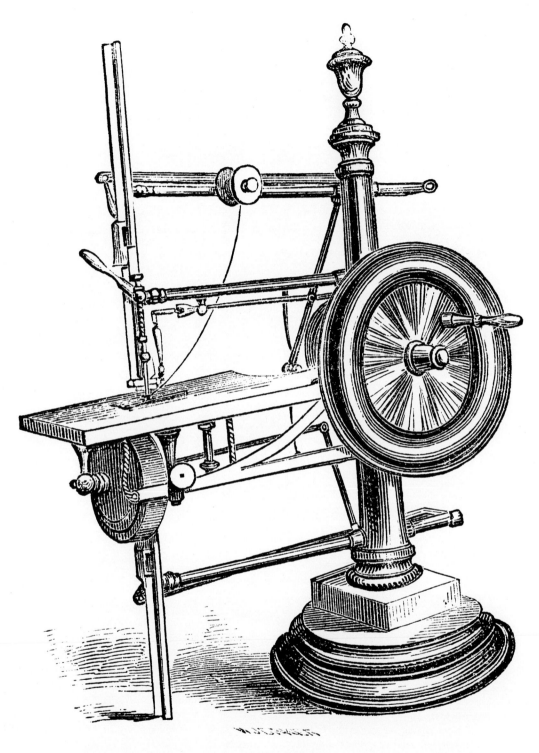

Figure 117.—From an advertising brochure, marked in ink, "The National Portrait Gallery, 1855," in the Singer Company's archives. The brochure states "Howard & Davis, 34 Water Street, Boston, Massachusetts Sole Manufacturers of Robinson's Patent Sewing Machine with Rope[r]'s Improvements." (Smithsonian photo 48091–F.)

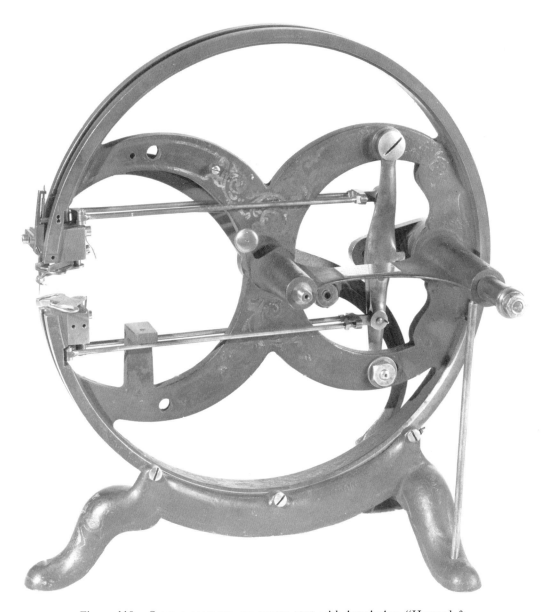

Figure 118.—Sewing machine of about 1856 with inscription "Howard & Davis Makers, Boston, Mass. Robinson & Roper Pat. Dec. 10, 1850, Aug. 15, 1854"; the drive wheel and the circular stitching plate of this machine are missing. (Smithsonian photo 48440–C.)

Figures 117 and 118.—Robinson and Roper sewing machines, 1855–1856. This is one of the few machines producing a backstitch or half backstitch to realize any commercial success. Manufactured a very short time by Howard & Davis, it was a short-thread machine, based on the Frederick Robinson patent of December 10, 1850, and the Samuel Roper patent of August 15, 1854. Roper produced additional improvements for which he received a patent on November 4, 1856. In the *Scientific American*, November 1, 1856, the new machine was discussed: "Robinson & Roper exhibit their new improved sewing machines, which appear to operate with great success. Two needles are employed, the points of which are furnished with hooks that alternately catch the thread and form the stitch. The finest kind of cotton thread or silk can be used. The work appears well done. Price $100."

Figure 119.—Illustrated page in a Shaw & Clark advertising brochure, published in late 1864. (Smithsonian photo 61321.)

Figure 120.—SHAW & CLARK SEWING MACHINE (Page patent) of 1867, Chicopee Falls, Massachusetts. (Smithsonian photo 48216–L.)

Figures 119 and 120.—SHAW & CLARK SEWING MACHINES. In addition to the early style Monitor sewing machine sold by Shaw & Clark without a name or any identifying marks, the company continued to manufacture machines after a lawsuit with the "Combination" forced them to take out a license. They manufactured an adapted version of their Monitor and an entirely new design patented in 1861. Their machines were now marked with the company name and a list of patent dates including those of Howe, Wheeler and Wilson, Grover and Baker, and Singer and the Batchelder patent, together with their own design patents. In 1867 the company moved from Biddeford, Maine, to Chicopee Falls, Massachusetts. In the same year, they began manufacturing a machine of the design patented by T. C. Page. The company is believed to have become the Chicopee Sewing Machine Company which appeared the following year and remained in business only a very short time. One Chicopee sewing machine is in the Smithsonian collection.

Figure 121.—Singer "Traverse Shuttle Machine—Letter A." (Smithsonian photo 58984.)

Figures 121 and 122.—Singer sewing machines. From 1850 to 1858 the Singer company produced heavy manufacturing-type sewing machines similar to the patent model shown earlier (fig. 28). The first machine for family use, Singer's new "Family" sewing machine (fig. 33) was manufactured from 1858–1861. Their second-style family machine was called the "Traverse Shuttle Machine—Letter A;" it was manufactured from 1859 to 1865, when they introduced their third family machine and called it the "New Family" sewing machine. This style machine continued until about 1883 when the "Improved Family" machine appeared. In addition to the lockstitch machines, Singer also manufactured chainstitch machines, and many highly specialized manufacturing machines.

From 1857 through the 1880s, the Singer machines were marked with two serial numbers. It is possible that the numbers were related to the "Combination" royalties paid by the Singer company. Until about 1873 there was a difference of exactly 4,000 in the two numbers, thus one machine would be marked 12163 and directly below it would be marked 16163. From 1873 the last three digits of the two numbers continued to be the same but the lower number might be much lower in value than either number used in earlier years. The larger number is believed to have been a record of total production while the lower number may have referred to a machine of a particular style. The Singer company records can shed no light on the meaning of the top (or lower of the two) serial numbers. Generally, in the earlier machines, the difference in the two numbers will not affect the dating of a machine by more than one year.

Since dating by serial number can only be estimated, the two numbers do not add an appreciable variable prior to 1873. Only the larger number, however, should be considered in dating machines after 1873.

Serial Number	Year	Serial Number	Year
1—100	1850	99427—123058	1864
101—900	1851	123059—149399	1865
901—1711	1852	149400—180360	1866
1712—2521	1853	180361—223414	1867
2522—3400	1854	223415—283044	1868
3401—4283	1855	283045—369826	1869
4284—6847	1856	369827—497660	1870
6848—10477	1857	497661—678921	1871
10478—14071	1858	678922—898680	1872
14072—25024	1859	898681—1121125	1873
25025—43000	1860	1121126—1362805	1874
43001—61000	1861	1362806—1612658	1875
61001—79396	1862	1612659—1874975	1876
79397—99426	1863		

Since records of annual production from 1877 to the turn of the century are not complete, it is difficult to establish yearly approximations. Using the machines submitted as patent models, and thus known to have been manufactured before the date of deposit, however, has provided us with the following date guides. By 1877 there had been 2 million machines manufactured, 3 million by 1880, 4 million by 1882, 5 million by 1884, 6 million by 1886, 7 million by 1888, 8 million by 1889, 9 million by 1890, and 10 million by 1891.

Figure 122.—Singer "New Family" sewing machine. (Smithsonian photo 58987.)

Figure 123.—STANDARD SEWING MACHINE of about 1870. This chainstitch machine is believed to have been made by the company that later became the Standard Shuttle Sewing Machine Company, when they began manufacturing lockstitch machines about 1874. This machine is marked with the name, "Standard," and with the dates "Patented July 14, 1870, Patented Jan. 22, 1856, Dec. 9, 1856, Dec. 12, 1865." The dates refer to the reissue and extended reissue of the Bachelder and the A. B. Wilson patents. The number of chainstitch machines of this type that were manufactured is not known. (Smithsonian photo 45506–C.)

Figure 124.—Taggart & Farr sewing machine, front view. (Smithsonian photo 48216–P.)

Figures 124 and 125.—Taggart & Farr sewing machine, 1860. The Taggart & Farr is an almost forgotten machine. It was based on Chester Farr's patent of August 9, 1859. The machine, however, was in commercial production as early as 1858, the year the patent application was made. Using two threads—both taken directly from the spool—to form a chainstitch, the machine was operated basically by treadle but also by hand. The drive wheel is missing on this machine, but it would normally appear on the right.

The name and patent date were painted on the end of the machine. This was true of many other machines of this period, which is why so many go unidentified once the paint has become worn. Several thousand Taggart & Farr machines were manufactured, but the company is believed to have had a short life, for it was among those that had disappeared by 1881.

Figure 125.—Taggart & Farr sewing machine, end view. (Smithsonian photo 48216–M.)

Figure 126. Watson's family sewing machine, as illustrated in *Scientific American*, December 13, 1856. The single-thread machine patented by William C. Watson, November 25, 1856. Watson had exhibited a "single-looping machine" as early as 1853 at the New York Industry of All Nations Exhibition; however, the earliest Watson machines were two-thread lockstitch machines, as described in the *Scientific American*, August 10, 1850. Although the magazine reported that the inventor had applied for a patent, the earliest lockstitch patent issued to William C. Watson was on March 11, 1856. A few of his machines were made in 1850, the article continued, "several of these machines are nearly finished . . . persons desirous of seeing them can be gratified by calling upon Messrs. Jones & Lee." A Watson machine was exhibited by Jones & Lee at the Sixth Exhibition of the Massachusetts Charitable Mechanics Association held in Boston in September 1850.

Included in the information provided in the 1856 *Scientific American* article was the proposed machines "of this character—8 inches by 5—for families at a retail price of $10." The less elaborate "Watson Ten Dollar Machine" was manufactured by Watson & Wooster from late 1856 to about 1860. Several of these are in the Smithsonian collection of patent models (see Herron's Patent, fig. 95), and at least two with their original paper labels exist, one in the Bennington Museum in Vermont and one in the Old Slater Mill Museum in Pawtucket, Rhode Island. No examples of this earlier, more elaborate, single-thread machine are known to have survived. (Smithsonian photo 48221–B.)

Figure 127.—West & Willson sewing machine of about 1859. The West & Willson machine, manufactured under the patent of H. B. West and H. F. Willson, enjoyed a very brief span of popularity. The patent covered the peculiar method of operating a spring-looper in combination with an eye-pointed needle to form a single chainstitch, but whether machines of this single-thread variety were manufactured is unknown. The machine illustrated here is a two-thread machine of basically the same description. It stitches from left to right and bears serial number 1544 and the inscription "West & Willson Co. patented June 29, 1858." (Smithsonian photo 49456–A.)

119

Figure 128.—Wheeler & Wilson sewing machine of about 1872. Serial number 670974. (Smithsonian photo P63149–A.)

Figure 129.—WHEELER AND WILSON No. 8 sewing machine of about 1876. (Smithsonian photo 17663-C.)

Figures 128 and 129.—WHEELER AND WILSON sewing machines. The Wheeler and Wilson company was the largest manufacturer of sewing machines in the 1850s and the 1860s.

From late 1851 to 1856 it was the Wheeler, Wilson, Co., Watertown, New York; and from 1856 to 1905, it was Wheeler & Wilson Mfg. Co., Bridgeport, Connecticut.

The style of the head changed very little during these years (see figs. 26 and 27). Both a table style with iron legs and a cabinet model were made: the head was usually mounted to stitch from left to right. In 1861, the company introduced the famous glass presser foot, patented on March 5 of that year by J. L. Hyde. The presser foot was made of metal but shaped like an open U into which was slid a small glass plate, with a hole for the needle descent. The glass allowed the seamstress to observe the stitching and to produce very close-edge stitching. It remained a favorite of many women for years. In 1876, the new No. 8 machine was introduced and a new series of serial numbers was initiated. It is, therefore, imperative to know that the machine is one of the earlier style machines before using the following list of serial numbers to date the machines, approximately as follows:

Serial Number	Year	Serial Number	Year
1—200	1851	141100—181161	1864
201—650	1852	181161—220318	1865
651—1449	1853	220319—270450	1866
1450—2205	1854	270451—308505	1867
2206—3376	1855	308506—357856	1868
3377—5586	1856	357857—436722	1869
5587—10177	1857	436723—519930	1870
10178—18155	1858	519931—648456	1871
18156—39461	1859	648457—822545	1872
39462—64563	1860	822546—941735	1873
64564—83119	1861	941736—1034563	1874
83120—111321	1862	1034564—1318303	1875
111322—141099	1863	1138304—1247300	1876

Records of the second series of serial numbers dating from 1876 are not available.

Figure 130.—WHITE SEWING MACHINE. Although the White sewing machines date from 1876, Thomas H. White had been busy in the manufacture of sewing machines for many years prior to this. White is known to have been associated with Barker in the manufacture of the Brattleboro machine and later with Grout in producing one of the several New England machines. In 1866 he moved to Cleveland, Ohio, and began manufacturing machines for sale under special trade names through selling organizations. In 1876, the White Sewing Machine Company was formed and machines were sold under the White name.

The machine illustrated is a standard lockstitch machine, which would have been set into a sewing-machine table and operated by a treadle. The small handle was used to start the wheel, and thus the stitching operation, in the forward direction. This machine bears the serial number 28241 and the following patents: "Mar. 14, 1876, May 2, 1876, Oct. 24, 1876, Jan. 16, 1877, Mar. 20, 1877, Mar. 27, 1877," which are primarily the patents of D'Arcy Porter and George W. Baker.

The machines of the 1870s may be dated approximately as follows:

Serial Number	Year
1—9000	1876
9000—27000	1877
27001—45000	1878
45001—63000	1879

(Smithsonian photo 48329–H.)

Figure 131.—WILLCOX AND GIBBS SEWING MACHINE, serial number 296572, of about 1878. From 1857 to the turn of the century, the style of the Willcox and Gibbs sewing machine changed very little (fig. 39). It was the most popular and the most reliable of the many chainstitch machines. In addition to the basic mechanical patents, Gibbs also patented the design of the sewing-machine head in 1860. In the specifications, he described it as an open ring set on a base or pedestal. The lower part of the open section supported the cloth plate. The design of the head, intentionally or not, formed a perfect letter G, the initial of the inventor. Later the machine head as a letter G was incorporated into the company's trademark. Additional patents were also granted to James Willcox for a leg and treadle design and to Charles Willcox for mechanical improvements.

It has not been possible to secure information on records of serial numbers from the late 1870s through the 1920s to aid in dating machines of that period. For the preceding years, however, the machines may be dated approximately as follows:

Serial Number	Year	Serial Number	Year
1—10000	1857	100001—115000	1867
10001—20000	1858	115001—130000	1868
20001—30000	1859	130001—145000	1869
30001—40000	1860	145001—160000	1870
40001—50000	1861	160001—190127	1871
50001—60000	1862	190128—223766	1872
60001—70000	1863	223767—239647	1873
70001—80000	1864	239648—253357	1874
80001—90000	1865	253358—267879	1875
90001—100000	1866	267880—279637	1876

Although the Willcox and Gibbs company continued in business until the 1970s, production in the last several decades was limited to specialized manufacturing machines rather than in family machines. (Smithsonian photo 58986.)

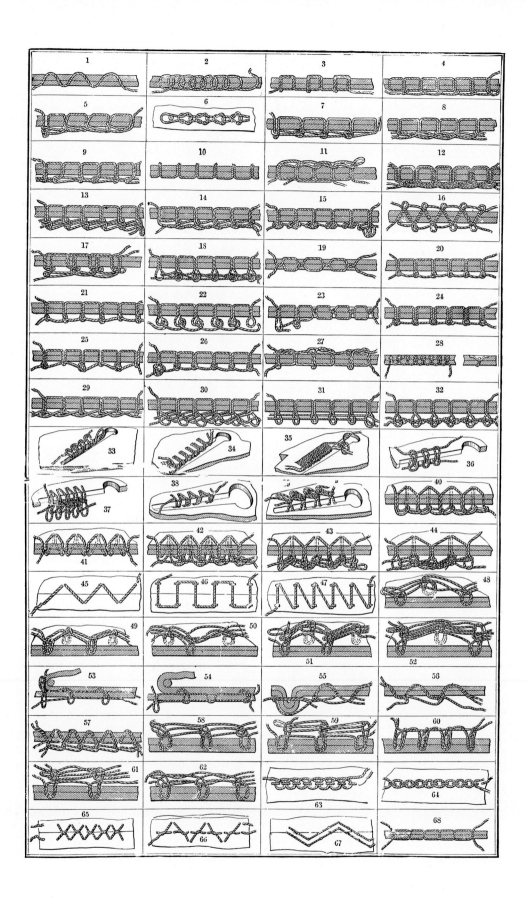

Figure 132.—ILLUSTRATION from *Knight's American Mechanical Dictionary*, vol. 3, p. 2122. The 68 sewing-machine stitches in use by 1882 are as follows:

SINGLE THREAD
1. Running stitch.
2. Back stitch.
3. Fast stitch.
4. Chainstitch.
5. Coiled-loop chainstitch.
6. Knitted-loop chainstitch.
7. Knotted-loop chainstitch.
8. Loop enchained by second alternate stitch.
9. Each loop locks and enchains alternate loops.
10. Staple stitch (for waxed threads only).

TWO THREADS
11. Double-needle chainstitch.
12. Double-thread chainstitch (one needle).
13. Double-looped chainstitch.
14. Chain with interlocking thread.
15. Under-thread through its own loop.
16. Two needles penetrate fabric from opposite sides.
17. Two needles working from the same side.
18. Double interlocking loop.
19. Lockstitch.
20. Twist in needle thread.
21. Double twist in needle thread.
22. Twist in shuttle thread.
23. Double twist in shuttle thread.
24. Knot stitch, shuttle thread knotted at every stitch.
25. Knot stitch, shuttle thread knotted at every other stitch.
26. Knot stitch, shuttle thread through the needle thread loop and knotted around the loop.
27. Shuttle thread pulled to the surface and interlocked with succeeding stitch to form an embroidery stitch.
28. Wire-lock stitch, thread locked in place with wire.

THREE THREADS
29. Two shuttles, each locking alternate loops.
30. Double loop with interlocking third thread.
31. Two shuttle threads, both locking each loop.
32. Two shuttle threads intertwining and locking each loop.
33. Single thread; loop of needle thread drawn up over the edge and locked by needle at its next descent.
34. Two threads; loops of needle thread, above and below, extend to the edge of the fabric, and are locked by shuttle thread.
35. Two threads; needle penetrates back from edge, its loop passed to and interlocked by the needle at its next descent over the edge, and this second needle-loop locked by shuttle thread.
36. Two threads; shuttle thread drawn up over the edge of the fabric to the line of the needle thread.
37. Two threads; needle loop through the fabric locked by needle loop over the edge and second loop locked by second thread.
38. Two threads; edge of fabric covered by shuttle thread.
39. Three threads; third thread laid around the stitch at the edge of the fabric.

ORNAMENTAL STITCHES
40. Zigzag; single thread chainstitch (4).
41. Zigzag; two-thread lockstitch (19).
42. Zigzag; two-thread chainstitch (13).
43. Zigzag; chain stitch with interlocking thread (14).
44. Zigzag; double loop with interlocking third thread (30).
45. Zigzag; running stitch (1).
46. Zigzag; two needles and shuttle.
47. Zigzag; variation of 46.
48–52. Zigzag stitches for sewing straw braid.
53–62. Straight straw-braid stitches.
63–67. Special embroidery stitches.
68. Saddler's stitch.

Figure 133.—Dorcas sewing machine was manufactured by the American Sewing Machine Company. This machine was awarded a diploma at the Seventh Exhibition of the Massachusetts Charitable Mechanics Association held in Boston in September 1853; by February 1856, it was listed as a "defunct" machine. The style is typical of the early, heavy, cast-iron ones (see figs. 28, 34, and 99). (Smithsonian photo, Warshaw Collection.)

Figure 134.—Wilson's patent sewing machine was manufactured by the Wheeler & Wilson Manufacturing Company. The illustration is from an advertisement of about 1854. Note that the machine could be purchased with an attractive table. Even from this early period, the advertising was directed to women: "Among the many advantages contained in this machine [is] the perfect simplicity and ease in operating them.... A lady of the most delicate constitution can use them with comparative ease...." (Smithsonian photo, Warshaw Collection.)

Figure 135.—The factory of the Grover, Baker & Company on Haymarket Square in Boston, Massachusetts, is illustrated in an 1853 advertising brochure for their sewing machines. The first sewing machines were built in small machine shops, but very early the success of the sewing machine was assured and factories were constructed to produce them in larger quantities. (Smithsonian photo 74-12053.)

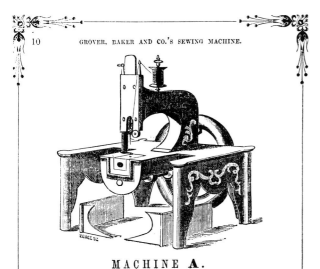

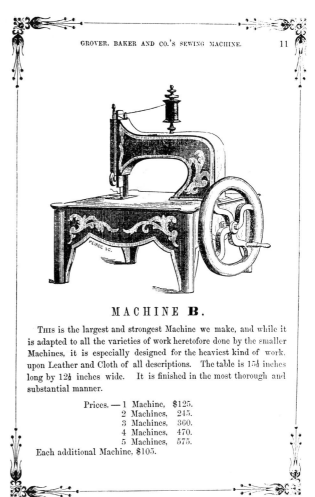

10 GROVER, BAKER AND CO.'S SEWING MACHINE.

MACHINE **A**.

THIS, although a little smaller than the B. Machine, is very strong and thoroughly constructed. It is adapted like that, to sewing upon heavy work of all kinds, both cloth and leather. It differs in construction from our other Machines, in that it is adapted to all work which can be done with them, and yet may also be readily applied to stitching round the tops of boots after they are closed and otherwise complete, and to all work requiring to be sewed in a cylindric form. When the brass slides are in, it is like the other Machines, and will do the same work; when they are out, as in the plate, it will be seen a cylinder is presented of convenient size for all work requiring that form of Machine. The table is 14¼ inches long, by 12½ inches wide.

Prices. — 1 Machine, $150.
2 Machines, 294.
3 Machines, 432.
4 Machines, 564.
5 Machines, 690.
Each additional Machine, $126.

Figure 136

GROVER, BAKER AND CO.'S SEWING MACHINE. 11

MACHINE **B**.

THIS is the largest and strongest Machine we make, and while it is adapted to all the varieties of work heretofore done by the smaller Machines, it is especially designed for the heaviest kind of work, upon Leather and Cloth of all descriptions. The table is 15½ inches long by 12½ inches wide. It is finished in the most thorough and substantial manner.

Prices. — 1 Machine, $125.
2 Machines, 245.
3 Machines, 360.
4 Machines, 470.
5 Machines, 575.
Each additional Machine, $105.

Figure 137

Figures 136–140.—GROVER, BAKER & COMPANY'S PATENT SEWING MACHINES included five styles in 1853. These machines varied slightly in style, size, and stitching capability with the price ranging from $60 for the smallest and simplest machine to $150 for the largest and strongest; prices were reduced when more than one machine was purchased. These machines were intended for manufacturing rather than home use. (Machine A, Smithsonian photo 74-10230; B, Smithsonian photo 74-10231; C, Smithsonian photo 74-10232; D, Smithsonian photo 74-10233; and E, Smithsonian photo 74-10234.)

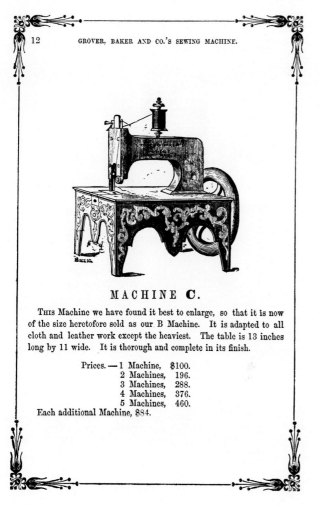

Figure 138

Figure 139

MACHINE C.

This Machine we have found it best to enlarge, so that it is now of the size heretofore sold as our B Machine. It is adapted to all cloth and leather work except the heaviest. The table is 13 inches long by 11 wide. It is thorough and complete in its finish.

Prices. — 1 Machine, $100.
2 Machines, 196.
3 Machines, 288.
4 Machines, 376.
5 Machines, 460.

Each additional Machine, $84.

MACHINE D.

This machine is a little smaller than the preceding, its table being 11¼ inches in length by 10 in width. It is well suited to sewing upon all cloth work except grain bags and heavy canvass. We have dispensed with the extra finish on this machine, but all the working parts are of the same perfect finish as the other machines.

Prices. — 1 Machine, $ 80.
2 Machines, 157.
3 Machines, 231.
4 Machines, 302.
5 Machines, 370.

Each additional machine $68.

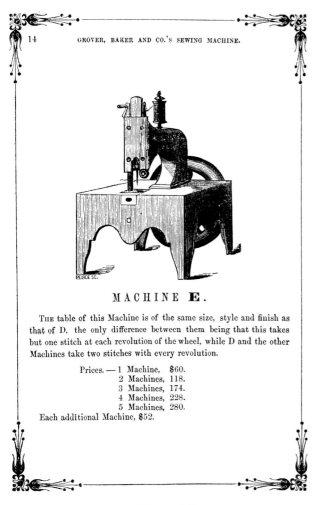

Figure 140

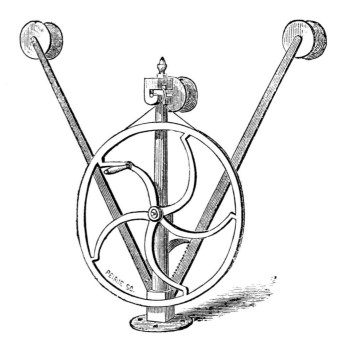

Figure 141.—A SPOOLING MACHINE is illustrated in the 1853 brochure entitled *Grover, Baker & Co's Patent Sewing Machine*. "This is a small but very convenient machine for spooling thread or silk. It is portable and will be found of much use to those using [sewing] machines or having occasion to wind thread of any kind. Price $3.50." It was especially essential for Grover, Baker & Company to offer a device of this type since their machine was a two-thread type that took both threads from a spool. Although some thread was sold on spools, thread was commonly sold in skeins in the mid-19th century, and a convenient implement to transfer the thread from skein form to spool form would have been most essential. (Smithsonian photo 74-10235.)

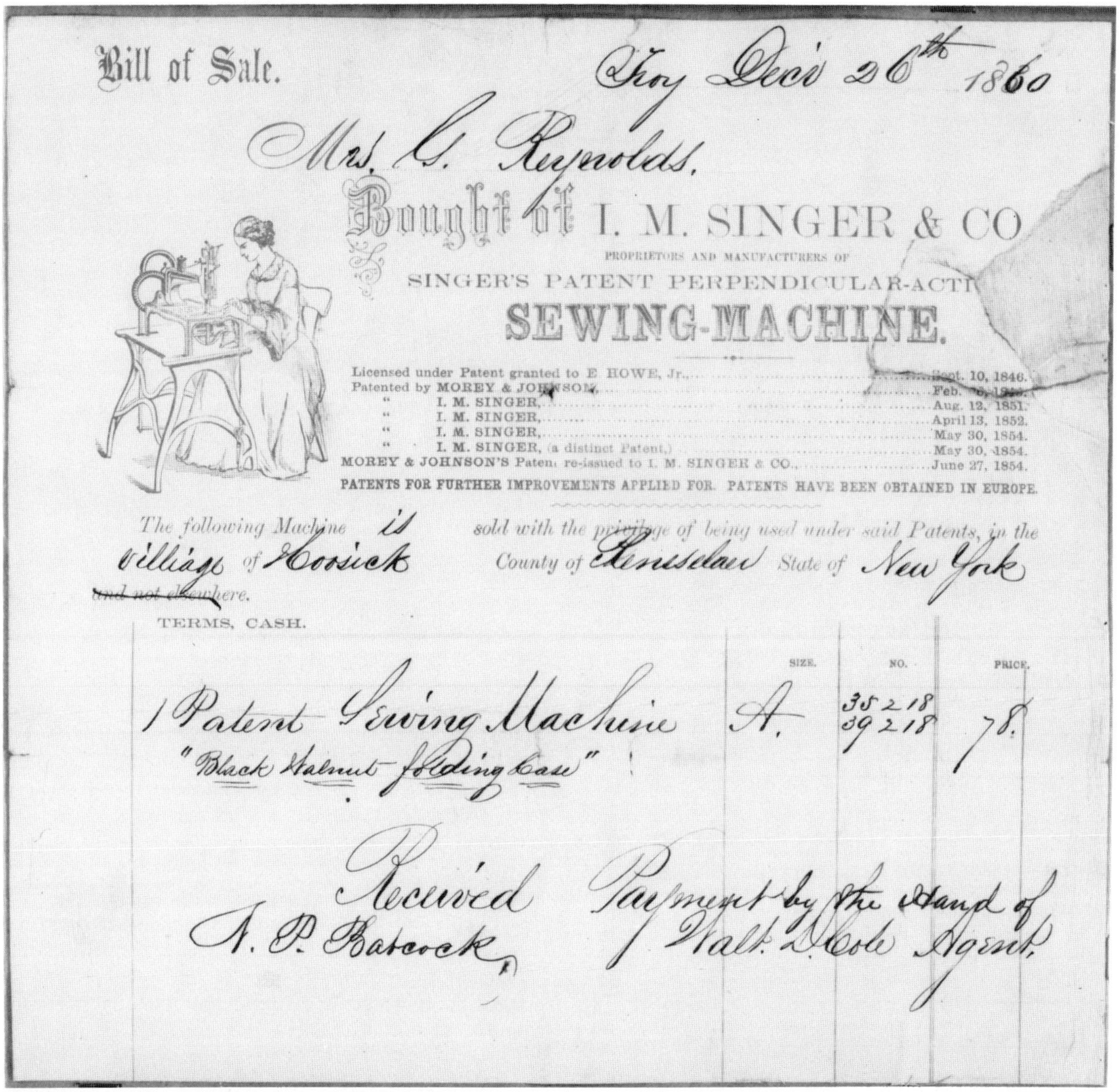

Figure 142.—THIS SINGER SEWING MACHINE BILL OF SALE is dated in handwriting, December 26, 1860; however, it is evident from the printed "185_" beneath the date that the same printed forms were used in the previous decade. The sewing machine that is illustrated is the basic Singer manufacturing machine. Compare it to those in figure 30 of 1856, and in figure 207 of 1924. The machine for which the receipt was made was for a "black walnut folding case," which would have been a style similar to figure 145. (Smithsonian photo, Warshaw Collection.)

Figure 143.—WILLCOX & GIBBS NOISELESS FAMILY SEWING MACHINE is illustrated in an advertising booklet of about 1864. By the mid-1860s the design of the iron framework begins to fill in with more detail. The Willcox & Gibbs sewing-machine head changes very little from the earliest commercial machines of 1857 (fig. 39) to this machine of about 1864 to those produced in the 1920s. (Smithsonian photo 67205.)

Figure 144.—GREENMAN AND TRUE SEWING MACHINE purchased about 1862 in Springfield, Massachusetts, for $80. The same light, ornamental ironwork in the table framework of the late 1850s continues into the 1860s (see fig. 101). Even when the iron legs are cast in the same design as another machine, the treadles differ. (Smithsonian photo 66602.)

Figure 145.—WHEELER & WILSON SEWING MACHINE in a designated "half-case," a top hinged to turn back on itself. This style is known to have been used as early as 1860. (Smithsonian photo 67213.)

Figure 146.—CHICOPEE SEWING MACHINE was manufactured in Chicopee Falls, Massachusetts, about 1869. Heavier iron castings began to appear toward the end of the decade. (Smithsonian photo 70013.)

Figure 147.—THE ELLIPTIC SEWING MACHINE was a product of the Elliptic company, but manufactured at the Wheeler & Wilson factory about 1867. It looked very much like the Wheeler & Wilson machines. This model sold for $55. (Smithsonian photo 67225.)

Figure 148.—The Secor lock-stitch shuttle sewing machine as illustrated in a broadside of about 1870. The name of the sewing-machine company cast into the treadle, or some other part of the framework, became increasingly common. (Smithsonian photo 67219.)

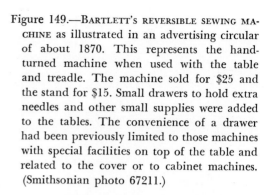

Figure 149.—Bartlett's reversible sewing machine as illustrated in an advertising circular of about 1870. This represents the hand-turned machine when used with the table and treadle. The machine sold for $25 and the stand for $15. Small drawers to hold extra needles and other small supplies were added to the tables. The convenience of a drawer had been previously limited to those machines with special facilities on top of the table and related to the cover or to cabinet machines. (Smithsonian photo 67211.)

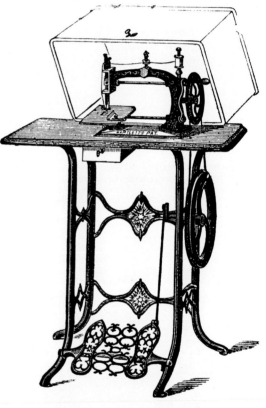

Figure 150.—The Home shuttle sewing machine sold by Johnson, Clark & Company about 1870. The iron legs are quite similar to the 1858 Ladd, Webster and Company machine (fig. 101), but heavier and with a less artistic flow. The treadle is ornate, and the table has a drawer. (Smithsonian photo 70014.)

Figure 151.—THE HOME SEWING MACHINE of about 1870 shows one of the many head styles that is quite similar to the Singer family machines. (Smithsonian photo 67203.)

Figure 152.—A SEWING MACHINE TABLE of about 1870 sold to be used with the Buckeye sewing machine (fig. 77). Hand-turned sewing machines were frequently sold with or without tables, to be used as a hand machine or a treadle machine. (Smithsonian photo 67214.)

Figure 153.—THE WEED SEWING MACHINE of about 1871 has the leg style of the earlier Wheeler & Wilson machines. (Smithsonian photo 67218.)

Figure 154.—The Singer "New Family" sewing machine in the simple table, one-drawer style sold for $60 in 1872. The one pictured in figure 122 is of the same era, but shows the stand with two drawers and a drop leaf, which would have been more expensive. (Smithsonian photo 67217.)

Figure 155.—The Remington sewing machine of about 1873 illustrated in an advertising circular. The sewing-machine heads of many of the family sewing machines of this period imitated the Singer machine. (Smithsonian photo 67228.)

Figure 156.—Wilson shuttle sewing machine of about 1873, illustrated with full cabinet and folding cover, which sold for $125; with a lift-off cover, the machine was priced at $110. The folding cover was hinged to lie flat. Ornate moldings were added to the cabinets of the early 1870s. (Smithsonian photo 67216.)

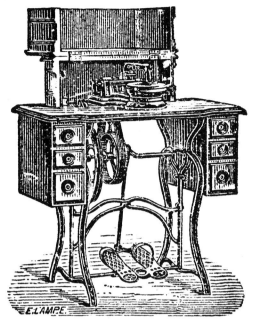

Figure 157.—FLORENCE SEWING MACHINE as illustrated in an 1873 leaflet, the *Florence Advocate*, showing the new patent "Bureau Case"; the machine and six-drawer table sold for $65. This basic design of three drawers on each side remained popular throughout the years of the treadle sewing machine. (Smithsonian photo 67207.)

Figure 159.—FLORENCE SEWING MACHINE in a full walnut cabinet sold for $100 in 1873. (Smithsonian photo 67227.)

Figure 158.—FLORENCE SEWING MACHINE illustrated in its simplest table form in 1873 sold for $45. Most of the Florence machines were sold as side-feed machines, but it was also mounted on the table to stitch from back to front and was termed the "Back Feed Machine." (Smithsonian photo 67223.)

Figure 160.—FLORENCE SEWING MACHINE with a hinged top and the familar ornamented legs, which were introduced about 1870, sold for $55 in 1873. (Smithsonian photo 67210.)

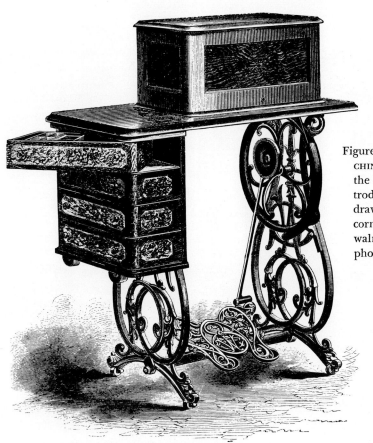

Figure 161.—WILLCOX & GIBBS FAMILY SEWING MACHINE with the usual style head, not shown, and the same ornamental iron work in the legs introduced a new "Patent Tray" in 1874. Four drawers are shown that pivot at the left front corner and swing out from the back. In black walnut, the machine sold for $100. (Smithsonian photo 67209.)

Figure 162.—DOMESTIC SEWING MACHINE with "folding case" and "fancy stand." The extended sewing surfaces at either side of the stationary head and the drop leaf seen at the rear of the stand are so hinged that they can be raised to form a boxlike covering over the head. In 1875, this machine in black walnut sold for $100 and in rosewood for $105. The legs are like those used by the Florence Company earlier in the decade. (Smithsonian photo.67224.)

Figure 163.—THE INDEPENDENT NOISELESS FAMILY SEWING MACHINE sold for $50 about 1875. The legs are cast in a design similar to the early Wheeler and Wilson machines. (Smithsonian photo 67221.)

Figure 164.—THE WILSON SHUTTLE SEWING MACHINE shown with the plain table sold for $50 about 1875. The iron framework is similar to those of the late 1850s, but it is slightly heavier and has lost its detail. (Smithsonian photo 67208.)

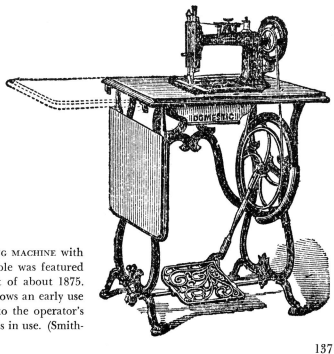

Figure 165.—THE DOMESTIC SEWING MACHINE with a "plain style" black walnut table was featured for $65 in an advertising leaflet of about 1875. This designated "plain" table shows an early use of the hinged leaf, which adds to the operator's work surface when the machine is in use. (Smithsonian photo 67215.)

Figure 166.—WILSON SEWING MACHINE as exhibited at the Centennial in 1876 represents the ultimate in sewing-machine-case ornamentation. Illustration from *Treasures of Art, Industry and Manufacture Represented in the American Centennial Exhibition, Philadelphia 1876,* Cosack & Company, publishers, Buffalo, New York, 1877, plate 34. (Smithsonian photo 76-1241.)

Figure 167.—THE CENTENNIAL SEWING MACHINE sold for $40 in 1876. The simple lines of the pleasantly designed table legs are more typical of the late 1850s than the mid-1870s. (Smithsonian photo 67212.)

Figure 168.—THE NEW·B HOWE SEWING MACHINE as illustrated in an instruction book of about 1877. Minor changes in machine frame and the addition of a stitch-size regulator distinguishes it from the Howe machine of 1870 (see fig. 98). (Smithsonian photo 67206.)

Figure 169.—DAVIS SEWING MACHINE COMPANY exhibit at the 1876 Centennial Exposition in Philadelphia. (Smithsonian photo 74-3806-3.)

Figure 170.—WILSON SEWING MACHINE COMPANY exhibit at the 1876 Centennial Exposition in Philadelphia. (Smithsonian photo 74-3806-12.)

Figure 171.—SEWING MACHINE with an Edison electric motor, 1880. (Smithsonian photo 73-6651.)

Figure 173.—MORRISON SEWING MACHINE, style no. 5, 1881. (Smithsonian photo 74-2092.)

Figure 172.—WHEELER & WILSON NO. 8 FAMILY SEWING MACHINE of about 1880. The No. 8 machine was introduced a few years earlier. (Smithsonian photo 67204.)

Figure 174.—QUEEN SEWING MACHINE, 1881. (Smithsonian photo 74-2091.)

Figure 176.—NEW STEWART SEWING MACHINE as shown in the *Sewing Machine Advance* of May 1881, page 70. (Smithsonian photo 74-2089.)

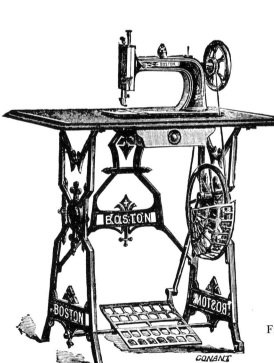

Figure 175.—BOSTON SEWING MACHINE, 1881. (Smithsonian photo 74-2090.)

141

Figure 178.—SINGER SEWING MACHINE, 1883, with drop-case open. (Smithsonian photo 74-2097.)

Figure 177.—SINGER SEWING MACHINE, 1883, with drop-case closed. (Smithsonian photo 74-2098.)

Figure 179.—New Wilson oscillating shuttle sewing machine (Wallingford, Connecticut), 1883. (Smithsonian photo 74-2095.)

Figure 181.—Crown sewing machine, 1883. (Smithsonian photo 74-2094.)

Figure 180.—Royal St. John sewing machine, 1883. (Smithsonian photo 74-2096.)

Figure 182.—Premium sewing machine, 1885. (Smithsonian photo 74-2099.)

Figure 183.—WHITE SEWING MACHINE in an elaborate cabinet that also served as a desk, 1888. (Smithsonian photo 73-2654.)

Figure 184.—WHITE SEWING MACHINE which could also be used as a desk. (Smithsonian photo 73-2653.)

Figure 185.—THE DAVIS COMPANY'S SEWING MACHINE with the same style cabinet as that of the White Company. The cabinets were made by an independent cabinetmaker in Cleveland. (Smithsonian photo 73-8013.)

Figure 186.—Head of the Splendid machine, 1888, shown in figure 187. (Smithsonian photo 73-8020.)

Figure 187.—Splendid sewing machine, 1888, made by the Parsons Manufacturing Company in Chicago. (Smithsonian photo 73-8019.)

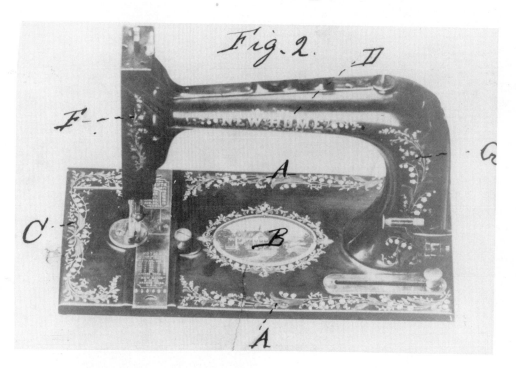

Figure 188.—Designs for sewing-machine decoration were also patented. This one is used on a New Home machine, patented May 20, 1884, by William Haebnel. The country house in the center was part of the design. (Smithsonian photo, Warshaw Collection.)

Figure 189.—Whitehill sewing machine, 1883. (Smithsonian photo 74-2093.)

Figure 190.—FAVORITE SEWING MACHINE, 1888, manufactured by the New Home Sewing Machine Company. (Smithsonian photo 73-8018.)

Figure 191.—WHEELER & WILSON SEWING MACHINE, 1889. (Smithsonian photo 73-8017.)

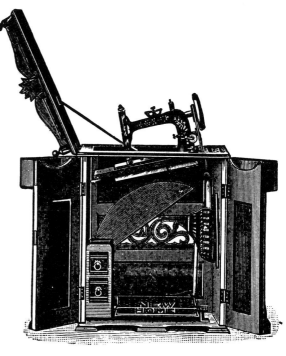

Figure 192.—NEW HOME DROP-CABINET, 1889, desk style. (Smithsonian photo 73-8016.)

147

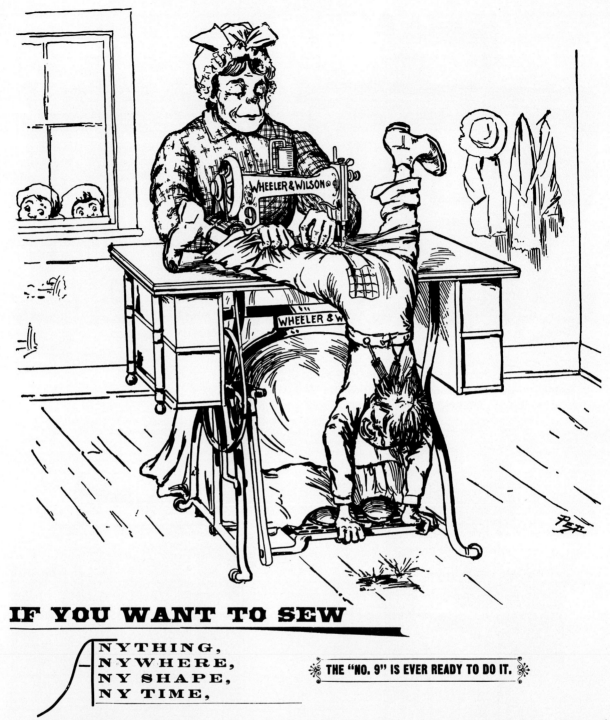

Figure 193.—WHEELER & WILSON SEWING MACHINE advertisement of 1892 in which new advertising techniques were being used. (Smithsonian photo 73-8012.)

Figure 194.—WHEELER & WILSON SEWING MACHINE of 1891 showing a later style addition in the form of a spindle detail in the cabinet (see fig. 191). (Smithsonian photo 74-2100.)

Figure 196.—NEW HOME SEWING MACHINE, 1896, showing the drop head. (Smithsonian photo 74-2103.)

Figure 195.—ADVANCE SEWING MACHINE, 1892, manufactured by the Davis Sewing Machine Company. (Smithsonian photo 74-2101.)

Figure 197.—WHITE SEWING MACHINE, 1892. (Smithsonian photo 74-2102.)

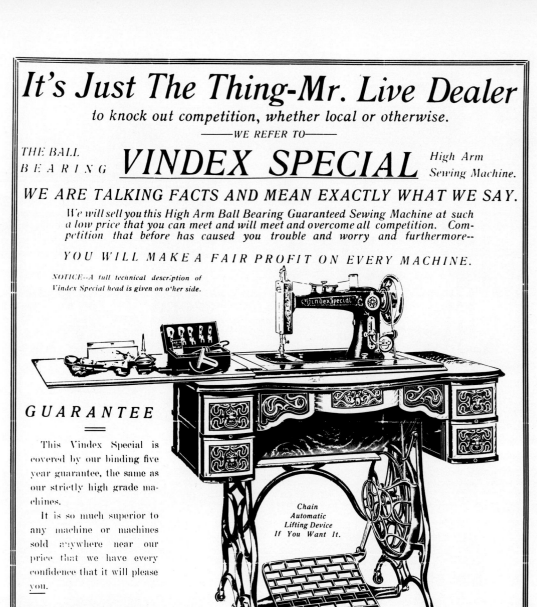

Figure 198.—Advertisement for the Vindex Special sewing machine which was manufactured about 1905. (Smithsonian photo, Warshaw Collection.)

Figure 199.—ANOTHER ADVERTISEMENT FOR THE VINDEX SPECIAL of about 1905. (Smithsonian photo, Warshaw Collection.)

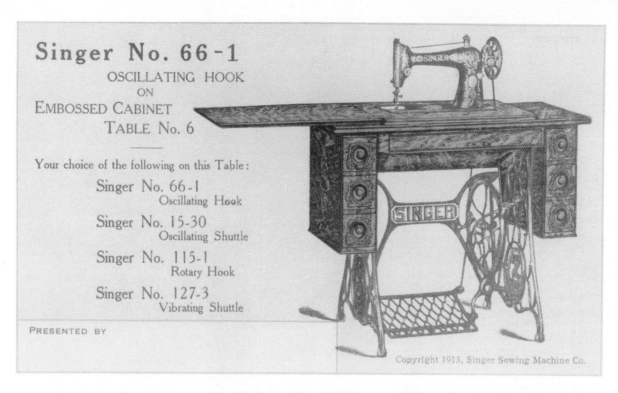

Figure 200.—SINGER FAMILY SEWING MACHINE of 1913. The embossed cabinet could be furnished with any one of four types of head. (Smithsonian photo, Warshaw Collection.)

Figure 201.—SINGER SEWING MACHINE of 1913 incorporated into a family scene and used in an advertisement. (Smithsonian photo, Warshaw Collection.)

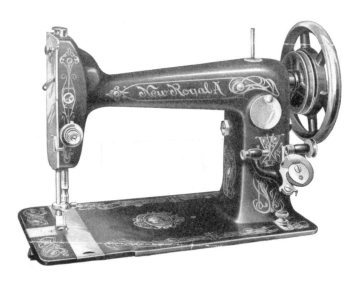

Figure 202.—Head of the New Royal sewing machine, 1913, manufactured by the Illinois Sewing Machine Company. (Smithsonian photo, Warshaw Collection.)

Figure 203.—New Royal sewing machine of 1913, manufactured by the Illinois Sewing Machine Company, showing the more elaborate of two of their cabinet styles. (Smithsonian photo, Warshaw Collection.)

Figure 204.—STANDARD SEWING MACHINE with an electric motor as it appeared in an advertisement in the March 1922 issue of the *Sewing Machine Times*. (Smithsonian photo 74-2121.)

Figure 205.—THIS STANDARD SEWING MACHINE of 1922 operated electrically. A neat console style evolves when the machine is closed. This style forecast the design of cabinets for several decades to come. (Smithsonian photo 74-2102.)

Figure 206.—THE STANDARD SEWING MACHINE COMPANY also continued to manufacture "Foot Power" machines in 1922. Note the motherly look of this woman, compared to the more "modern" woman using the electric machine in figure 204. (Smithsonian photo 74-4193.)

Figure 207.—SINGER SEWING MACHINE, 1924, manufacturing-machine class 1W, curved needle, is basically the Wheeler & Wilson sewing machine as manufactured by the Singer Company in the 1920s. It was designated "for fine lock stitching in the manufacture of shirts, collars, cuffs, etc." The machine head looks very much as it did in the 1860s. (Smithsonian photo 74-2123.)

Figure 208.—SINGER SEWING MACHINE manufacturing machines of the earliest style continued to be produced in 1924. (Smithsonian photo 74-2122.)

Machines for Factory Use

Figure 209.—WILLCOX & GIBBS FACTORY MACHINE for pleating and ruffling of 1874. Patent No. 156,119 was issued to Samuel Barney. (Smithsonian photo 73-11196.)

Figure 210.—FACTORY MACHINE for stitching stays in corsets, 1877. Patent No. 192,568 was issued to E. Corbett and C. F. Harlow. (Smithsonian photo 73-11198.)

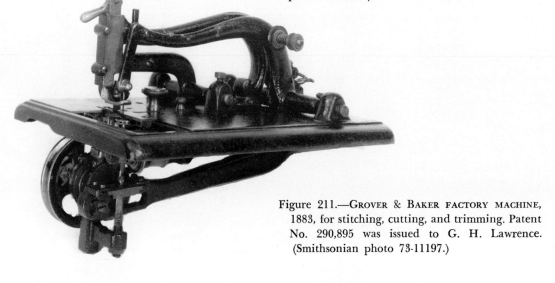

Figure 211.—GROVER & BAKER FACTORY MACHINE, 1883, for stitching, cutting, and trimming. Patent No. 290,895 was issued to G. H. Lawrence. (Smithsonian photo 73-11197.)

Figure 212.—SINGER FACTORY MACHINE of 1889 for making buttonholes has a device for cutting the buttonhole. Added in an 1890 patent is a mechanism for sewing on 4-eyed buttons for which Patent No. 425,422 was issued to J. P. Hallenbeck. (Smithsonian photo 73-11195.)

Figure 213.—WHEELER & WILSON FACTORY MACHINE of 1891 for a 3-stitch zigzag. (Smithsonian photo 73-11200.)

Figure 214.—STANDARD FACTORY MACHINE of 1892 for parallel rows of stitching, most widely used on men's shirts, for which Patent No. 479,369 was issued to Ferdinand Kern. (Smithsonian photo 73-11199.)

III. Sewing-Machine Styles

To a limited degree, the style of the sewing machine may be used to establish an approximate date. Since there are many exceptions, extreme caution must be used in this method of dating. Some of the styles of sewing-machine heads continued to be used for many decades. Minor improvements that were patented can be very helpful in limiting the date as these parts are frequently stamped with this patent information. Furniture styles of tables or cabinets also changed with trends for certain woods, manufacturing processes, and design taste.

1850s

Although a few sewing machines were sold in the late 1840s, the manufacture of sewing machines in quantity dates from 1851. In the early 1850s most of the machines were built to be used in manufacturing processes, although this might be on a small scale and sometimes even in the home. Since these machines were to be in constant use they were built rugged, usually of cast iron. The packing box might double as the table or a sturdy table might be furnished with the machine (see figs. 30, 31, 70, and 72). The exception to the ruggedly built machine was the light machine made by Wheeler and Wilson. Almost immediately, this company recognized that there was a large commercial market in filling each home with a family sewing machine. Although light in style, this machine became the shirtmakers' most valuable tool. By the middle 1850s both the Grover and Baker company and the Singer company had also introduced lighter weight machines for family use. By the end of the decade, a number of companies were producing machines for the home. The different machines made by these companies usually had a wood-top table with cast-iron legs; the casting was light in design and artistically attractive (see figs. 101, 105, and 127).

Cabinet machines appeared as early as the table models. The cabinets were usually of solid wood, walnut, mahogany, or rosewood, and simple in line (see figs. 35 and 116). The extreme popularity and desirability of the sewing machine as a consumer product is evident in the influx of inexpensive machines that appeared on the market as have been discussed in chapter four. By the end of the decade, the *Scientific American* reports in the July 30, 1859, issue that there were twenty-five sewing machine manufacturers in the United States making thirty different sewing machines of which 1,500 were being sold weekly. They also estimated that 100,000 sewing machines were in use at that time. The sewing machine had made a good beginning.

1860s

The light design of the ironwork of the 1850s gradually became more detailed and beautifully ornate. Unfortunately, by the end of the decade the castings had become heavy. Cabinet styles continued to be quite similar to those of the earlier decade. This is really a period of continuation for the styles were considered new and much of the market was still untouched.

1870s

From the late 1860s, probably due to the exhibition of sewing machines at large international expositions such as the one held in Paris in 1867, sewing-machine cabinets became more and more elaborate for those who could afford to pay for the ornate woodwork. It was understandable that, once the sewing machine was accepted as the mechanical wonder that it was, each manufacturer would attempt to make the packaging of his machine—the stand or cabinet—as eye-catching as possible. The 1870s was also the decade of many new innovations in styles and methods of adding drawers or drop leaves to the less costly table or stand.

The American sewing machine had won international favor at the London exhibition in 1862, the Paris one in 1867, and the Vienna exhibition in 1873. With this it had not only become an important export item around the world, but it was also widely copied by foreign manufacturers. At the 1876 Centennial exhibition in Philadelphia there were thirty American exhibitors of sewing machines in Machinery Hall. Decorated marquees housed the machines, and show cases exhibited examples of the work that the machines could produce. The Singer company, the largest sewing-machine manufacturer at this time, housed its exhibit in a building of its own. The major patents expired the year after the Centennial, and many new companies blossomed into existence. Some of these were companies that manufactured specialized sewing machines for very limited manufacturing tasks.

1880s

The motor could have been the new addition to the sewing machine in 1880. Although it is evident that it was possible this early, the practical application of electricity to operate the sewing machine did not come into use until electricity was in the home. Electric sewing machines were still several decades away. Reportedly, new ideas of lowering heads into cabinets were advertised. Comparison with figure 116 makes it apparent that the idea was not so new. From the 1880s on, more companies did use this method of lowering the machine to provide a flat surface when the machine was not in use; however, machines continued to be manufactured that required covering the machine head with a box. Oak became an increasingly important wood used in sewing-machine tables and cabinets. By the end of the decade, more and more specialized machines were manufactured each year by such firms as Union Special Machine Company and Merrow Machine Company.

1890s

New sewing-machine names on the market came to mean new styles from an existing company rather than a new company. The drop-head was still considered "new" in some advertisements, and the styles reflect the everyday domestic furniture of the era. The bentwood—as opposed to being cut into shape—technique was finally applied to sewing-machine head covers. Manufacturing machines had become industrial in nature as the specialization of each type of machine became more and more refined.

STYLES IN THE EARLY DECADES OF THE 20TH CENTURY

Embossed wood designs had become an important feature of the cabinets of sewing machines in the first decade. The cabinets themselves are most frequently the "bureau" style introduced earlier rather than the the full floor cabinet, although the latter is still found. Electric machines were being offered by most of the major companies by the second decade; however, the production of treadle machines continued to be the more popular for obvious reasons—lower cost and lack of electricity in the home. Hand-turned machines were manufactured by some of the companies, but these were primarily for export. They were not popular in America except for the children's variety. So-called "period style" cabinets appear on the market beginning in the mid-20s. This coincides with the revival of interest in period-style furniture, which sewing-machine manufacturers copied. Portable machines, which had not been popular since the earliest years, became increasingly so after electric machines and electricity were common. A table to house the treadle was no longer a necessity and any household table could serve the purpose; it was less important for the sewing machine to be a piece of furniture.

SEWING-MACHINE COMPANIES

Fanciful names for home sewing machines began to be used in greater numbers in the late 19th century, as noted by such appellations as Minnehaha, Splendid, Queen City, and others. For the most part, each new name represented a new company with a different machine, although the question of "difference" could be debated as many of the machines after 1877 looked alike. The smaller companies, especially the new ones, copied the style of the well-known, established machines. The idea of colorful names was not new in the 20th century, but increasingly became the rule rather than the exception. It was in the last decade of the 19th century that companies—like the newly formed National Sewing Machine Company—began selling their machines in increasing numbers to department stores, catalog houses, and other retail outlets, furnishing them with any name selected by the retailer. Sewing machines were even offered as premiums with newspaper subscriptions, and the name of the newspaper would be applied to the machine. Disappearing was the tradition that the machine must bear the inventor's or manufacturer's name. By the early 20th century, literally thousands of different-named machines were manufactured by less than a dozen sewing-machine companies in business at the time. Two companies did not follow this fashion, Singer and Willcox and Gibbs. The Willcox and Gibbs family machine continued to be the simple chainstitch. The "Singer" name reigned supreme, and it was the only one under which that company's machines were sold. Both the style of the Singer machine and the design of the trademark were widely copied by many companies both at home and abroad. Other companies also sold under their own name, but also added to the vast array of new names. Although sewing machines were not as decorative as they had been in the earlier period, verbally they were more colorful.

A brief history of each of the companies producing the sewing machines that were involved in this commercial name-calling will be given to help establish the dates of these machines. Company records were not kept as to which names were furnished to whom. Some helpful records exist, since sewing machines were built to last for years, and "parts suppliers" for these machines grew to be a respectable business in itself. To offer the correct parts, the parts companies had to inform the potential customer, and themselves, with the name of the manufacturer of the machine; the name stenciled on the machine was not sufficient for a part replacement. Two of the most prominent of these parts companies printed catalogs that have been very helpful. One is the *Sewing Machine Supplies and Parts*, catalog of the A. G. Brewer Sewing Machine & Supply Company, Chicago, Illinois, whose first edition was published in 1926. In the introduction, the manager of the company writes: "In getting this book out we endeavored to list the items now in demand by the Sewing Machine dealer, both old as well as the latest." The second is *Bryson's Verified Complete List of All Interchangeable Parts and Needles*, catalog of the C. M. Bryson Company of Cleveland, Ohio, which was copyrighted in 1933. The C. M. Bryson Company had been a sewing-machine sales agency in the late 19th century and became a parts supplier in the 20th century. The company slogan was, "If Bryson does not keep it, you can't get it."

BRIEF COMPANY HISTORIES

Davis Sewing Machine Company. First established in Watertown, New York, this company moved to Dayton, Ohio, in 1889–1890. The reasons given for the move were "first, the pressing necessity of increasing the manufacturing facilities of the company to meet the growing demand for their machines, and next, the great advantages offered for manufacturing in the West. . . ." Large factory buildings were constructed in Dayton and business flourished for more than thirty years. Although the name Davis Sewing Machine Company was retained, the actual manufacture of machines was taken over, at some time before 1924, by the H. M. Huffman Mfg. Co. On December 24, 1924, the Dayton operation ceased as the "National Sewing Machine Co., Belvidere, Ill. purchased good will, trade names and right to manufacture sewing machines under name Davis and all other names previously owned by Davis S. M. Co."

Domestic Sewing Machine Company. The Domestic machines were first manufactured in Norwalk, Ohio. The manufacturing rights were sold to a company in Newark, New Jersey, about 1871; the company was under contract to sell all the machines that they produced to the Domestic Sewing Machine Company in Ohio. In 1896 the two companies in New Jersey and Ohio were consolidated, although the factories continued to be located in New Jersey until about 1906. Between that date and 1914, Domestic machines were built in Buffalo, New York. In 1922 a New York City office for the Domestic Sewing Machine Company was opened. In 1924 the Domestic company became a wholly owned subsidiary of the White Sewing Machine Company and was located in Cleveland, Ohio. Domestic electric machines for home use were still being manufactured in the Cleveland factory in 1974.

A. G. Mason Manufacturing Company. A. G. Mason had been a general agent in Cleveland for the Davis Sewing Machine Company until 1903 when he began to manufacture sewing machines under his own name. He concentrated on building up sewing-machine departments in large retail stores. He gave particular attention in his manufacture of sewing machines to producing medium-priced, high-quality machines sold under special names chosen by the retail dealer. A. G. Mason died in 1916, and the company became a subsidiary of the Domestic Sewing Machine Company.

Household Sewing Machine Company. Beginning as the Providence Tool Company manufacturing household sewing machines, the company name was changed to Household Sewing Machine Company in 1890 and manufacture continued in Providence, Rhode Island, until 1906. Several other machine names were used for their products in addition to "Household."

Foley & Williams Manufacturing Company. This company developed about 1885 from the earlier H. B. Goodrich Company of Chicago, Illinois. The Foley and Williams company remained in business through 1924, but in the 1926 Brewer catalog they are listed as "out-of-business." Shortly after this, the company was reorganized as the Goodrich Sewing Machine Company and continued in active business through the mid-1930s.

New Leader Sewing Machine Company. It is not known whether this company was related to the earlier Leader company of Springfield, Massachusetts, which moved to Cleveland before it went out of business in 1884. The earliest record of the New Leader is 1926 and the company was listed as being in operation as late as 1933. The names given its sewing machines are related to the company name, e.g., New Century Leader.

Standard Sewing Machine Company. Frank Mack and William A. Mack, with years of experience in the sewing-machine business, began their own manufacturing as the Standard Sewing Machine Company, Cleveland, Ohio, in 1884. They both continued with the company until 1893. The Standard company continued to manufacture sewing machines in Cleveland through the early decades of the 20th century. Both electric and treadle machines were advertised in the early 1920s. The Standard Sewing Machine Company was sold to the OSAAN Fur Machine Company which was acquired by the Singer company in 1931.

Free Sewing Machine Company. This company, located in Rockford, Illinois, was an outgrowth of the Royal Sewing Machine Company that built a new factory in Rockford in 1890. By the end of December 1894, the Royal company had sold out to Gilbert Woodruff. In 1895 the Illinois Sewing Machine Company was founded using the remaining assets of the earlier company. Will C. Free joined the company in 1898 and became the president in 1910. He retained the Illinois Sewing Machine Company as a subsidiary, but organized a parent company—the Free Sewing Machine Company. In 1917, working with Westinghouse, an electric-model sewing machine was introduced under the name "Free-Westinghouse." The company continued in business in Rockford until 1958 when it moved to Los Angeles, California. Manufacture of sewing machines was discontinued in 1969.

National Sewing Machine Company. This company was formed in 1890 as the result of a consolidation of two older companies, the June Manufacturing Company and the Eldredge Sewing Machine Company both of which had been organized in 1879 after the expiration of the Sewing Machine Combination. Both companies had moved their factories from Chicago to Belvidere, Illinois, in 1886. The National Sewing Machine Company sold most of their machines through department stores and mail-order houses; Marshall Field's department store in Chicago began to sell National (Eldredge-built) machines in 1888, John Wanamaker's of Philadelphia purchased its first machine in 1892, and R. H. Macy's in New York in 1897. Eldredge-built machines were first sold to the Montgomery Ward company in 1889 and National continued to furnish them sewing machines. National produced its first electrical machines in 1917, which were sold through the Western Electric Company; their first so-called "period style" consoles were also introduced in 1917. On September 1, 1953, the National Sewing Machine Company was merged with the Free Sewing Machine Company as a wholly owned subsidiary.

New Home Sewing Machine Company. In 1862 Andrew Clark and William P. Barker, under the name of Clark and Barker, began manufacturing New England single-thread hand-turned sewing machines in Orange, Massachusetts. In 1865 Clark bought out Barker and continued in business for several years. A new firm was organized by Clark in 1867 and named the A. F. Johnson and Company. Valuable patents were purchased from Mr. Johnson and incorporated into the machines. The factories were enlarged and manufacture of the Gold Medal and Home Shuttle sewing machines were added to the New England machines. In 1869 the company was again reorganized, this time called the Gold Medal Sewing Machine Company. The manufacture of the Home sewing machine began in 1870; in 1877 when all the major patents—held by the Sewing-Machine Combination—expired, the New Home machine was developed by W. L. Grout, who was the company superintendent and general manager. The new machine was popular, and in 1882 the company name was changed to the New Home Sewing Machine Company. This company continued to manufacture sewing machines in Orange, Massachusetts, selling them under a wide variety of names. In 1930 they became affiliated with the Free Sewing Machine Company and moved to Rockford, Illinois. The company moved, with the Free company, to Los Angeles in 1958. In 1960 the New Home business was purchased by the Janome Sewing Machine Company, Ltd., of Tokyo, Japan, with offices in Santa Monica, California. To date, new Home sewing machines of Japanese manufacture continue to be built.

White Sewing Machine Company. Thomas H. White manufactured sewing machines in Templeton, Massachusetts, and in Orange, Massachusetts, for several years before he moved to Cleveland, Ohio, in 1866. At that time, he organized the White Manufacturing Company and built sewing machines for sales organizations bearing their trade names. In 1876 the company was organized into the White Sewing Machine Company and, for the first time, sewing machines were sold under the name "White." By 1919, 7 percent of the annual production was in the newly introduced electric machines. In 1924 the company acquired the Domestic Sewing Machine Company and the King Sewing Machine Company of Buffalo. The King company had been a subsidiary of Sears, Roebuck and Company and furnished machines to them. In 1925 the companies were consolidated to form the White Sewing Machine Corporation. With this, came a ten-year contract with Sears, Roebuck and Company to furnish their machines for a ten-year period. The White company continued to manufacture sewing machines in America through the 1960s. By 1974 the company had become the White Consolidated Industries with the White Sewing Machine Company, a distributor for White machines manufactured in Japan. The Domestic Sewing Machine Company remained a subsidiary and continued to manufacture electric machines for home use in Cleveland, Ohio.

Willcox and Gibbs Sewing Machine Company. The Willcox and Gibbs company of New York City was one of the few from the 1850s—the first decade of sewing-machine manufacture—to survive to the 20th century. The family chainstitch machine that had been the company's strong seller was phased out of the active selling program in 1926–1927; however, production of this machine, which had changed very little in style from that of the 1860s and 1870s, was continued intermittently with the last lot being produced in 1946. The machines were always sold under the company name, although this style machine was copied by several other companies and sold under their names. Beginning in 1875, the Willcox and Gibbs company began to include manufacturing machines in their line of production. The first one was the Chainstitch Visible Stitch Straw Hat Sewing Machine. It is believed to have been the first machine for stitching straw braids for the making of hats. For at least sixty years after its introduction, thousands of these machines were sold to the straw-hat industry throughout the world. This one was just the first of a long line of manufacturing machines which they continued to manufacture. These machines included lockstitch ruffling machines, lockstitch long-arm tucking machines, zigzag machines, overlock machines, hosiery welting and trimming machines, feldlock machines, and the American version of the Cornely embroidery machines that were based on the Bonnaz patent. Embroidery sewing machines were their forte and one of their more unusual machines was the Uniart embroidery machine, which could make a stitch as long as ¾-inch and duplicate the diagonal stitching found only in hand embroidery. Shell scalloping machines and bag machines were also among their long list of specialized sewing machines. The Willcox and Gibbs company discontinued manufacture in 1973.

Singer Manufacturing Company. The Singer company had been one of the three successful ones from the early 1850s. By the late 1860s, it finally surpassed the Wheeler and Wilson company in production and sales; in 1905–1907 Singer bought out the company. With this acquisition, they added the manufacture of a rotary-bobbin machine to their vibrating shuttle machines. Singer chainstitch machines had been introduced in the last quarter of the 19th century and continued to be popular into the second and third decade of the 20th century. From the beginning, Singer produced heavy-duty manufacturing machines, and family machines were introduced after a few years; both types of machines continued to be produced, the Singer name proudly displayed on each and every one. By the end of the 19th century, the Singer name was known around the world. A Singer manufacturing plant was located in Scotland, but agents were everywhere. Updated cabinets changed with the style and attempts to introduce an electric machine as early as 1889 are credited to the Singer company, but until electricity was commonly found in the household there was no great demand for the machine. The manufacturing machines were designed with special driving attachments so that they could be belted to the main source of power, whatever that might be. The wide variety of manufacturing machines included machines for harness and leather work, machines with two needles and two shuttles for canvas belting, bag machines, hat machines, overedge machines, buttonhole machines, and automatic button sewers. There were machines to stitch tents and wagon covers, to sew elastic banding, glove machines, special machines for darning and patching laundry nets, book and pamphlet machines, hemstitch machines, and long-arm and short-arm machines. The variety is endless. The major factories in the early decades of the 20th century were in Elizabeth, New Jersey, and Bridgeport, Connecticut, which was the old Wheeler and Wilson factory. The "Class 1w" machine produced in Bridgeport as late as 1925 was the same style as the Wheeler and Wilson sewing machine manufactured as early as the 1860s. It was still recommended for "fine lock stitching in the manufacture of shirts, collars, cuffs . . . ." The Singer company remains a very active one. Although many of their current machines are manufactured in foreign plants, some are still produced in the United States.

INDUSTRIAL SEWING MACHINES

From the beginning, sewing machine companies produced heavy-duty machines for manufacturing purposes. These machines were frequently similar in style and type to the light-weight family machines, but constructed to withstand constant daily use. As the specialized stitching potential of the sewing machine continued to expand, and after the basic patents had expired, companies were organized that confined their interests to the manufacture of sewing machines for industrial use. These machines became exceedingly specialized, some to the extent that only a few of one type were ever made. The machines were engineered to do a specific task for a specific manufacturer. This phase of industrial manufacture, as contrasted with the sewing machine as a consumer product, continues to thrive in the United States. Over fifty companies are actively engaged in producing sewing machines of all types essential to modern industry. Two of these companies began in the 19th century and have continued in business for almost one hundred years.

Union Special Machine Company. This company started in 1880 when R. G. Woodward was employed to build the improved bag machine originated by Jasper W. Corey. Called the Union Bag Machine, it reportedly "will last from two to three years in constant use run at a speed of 800 to 1000 stitches per minute." Corey had worked in the sewing-machine repair shop of Lorenz Muther who joined the Union Bag Machine Company at its inception in 1881. He designed the two-needle vamping machines used extensively in the manufacture of shoes in the early part of the 20th century. Another early contribution was made by John W. Dewees, the inventor of the trimming device—patented October 31, 1882—applied to the Union machines. Dewees' trimmer was considered the most effective and durable device introduced for this type of work. Lansing Onderdonk, who patented special attachments such as one for ruffling in 1883, joined the company now known as the Union Special Machine Company in 1888. He proceeded to make special machines for stitching carpets, a cylinder vamping machine for the shoe trade, and machines to work at higher speeds. By 1907 they had machines that were capable of attaining up to 4,000 revolutions per minute. Many of the machines in the 20th century were overedge machines related to the hosiery and knitted-wear industry and machines for specialized tasks in the clothing industry such as belt-loom machines that folded, stitched, cut, and counted the belts; the machine for setting in gathered sleeves; and one that stitched in the neckband facing in one operation. Bag and filled-bag closing machines continued to be a specialty of the company. The Union Special Machine Company continues to fill the needs of industry today.

Merrow Machine Company. In 1877 J. M. Merrow patented a machine in the class called "crochet," because it imitated the handwork done on the edges of fabric, especially knitted ones. Later, the machine was improved and placed on the market in 1889 by the Merrow Machine Company. Upon these fundamental machines, numerous additions and improvements were made in the 20th century. In the mid-1920s, they were making a variety of high-speed machines. The notable types were two-thread plain-crochet machines, single-thread blanket-hemming machines, two-thread shell-stitch machines, two-thread overedge machines, and two- and three-thread trimming and overseaming machines. The company still is in active operation in Hartford, Connecticut.

IV. 20th-Century American Sewing Machines

Sewing-Machine Name	Sewing-Machine Company	Sewing-Machine Name	Sewing-Machine Company
Abbott	National	Alberta Special	A. G. Mason Mfg. Co.
A.B.C.	A.B.C. Machine Co. (1906—)	Albion	National
		Alde	Standard
Abelone Furniture Co.	A. G. Mason Mfg. Co.	Alden	Standard
Abendale	National	Alexander	A. G. Mason Mfg. Co.
Abenschule	National	Alexandria	Davis
Aberdeen	Davis	Alfords	Davis
Abilene	White	Alladin	National
Abney Machine Co.	A. G. Mason Mfg. Co.	Allday	A. G. Mason Mfg. Co.
Abrams Special	White	Allens Queen	A. G. Mason Mfg. Co.
Absequis de la Colonia	Goodrich	Alleys Special	Free
Abstbuck Special	New Home	Alliance	Davis
A. B. and W.	Davis	Alliance	National
Acadia	Standard	Allison	White
Acme	Davis	Allison Special	Free
Acme	National	Alma	National
Acme	Standard	Almacen San Miguel	National
Acme Rotary	National	Aloha	Davis
Acrumann	A. G. Mason Mfg. Co.	Alosh	Goodrich
Acworth	National	Alpha	White
Adams	Davis	Al Ponetics de Liverpool	National
Adams	White	Alston	Standard
Adarondack	A. G. Mason Mfg. Co.	Alta	National
Adelaide	Davis	Alto	National
Adkins	A. G. Mason Mfg. Co.	Alva	National
Adla	Goodrich	Always-Good	Davis
Adlake	Goodrich	Always Ready	National
Admiral	Davis	Alzira	White
Adrian Hardware Co.	A. G. Mason Mfg. Co.	Amazon	National
Advance	Davis	Amena	Standard
Aero	Standard	America	Standard
Aetna New	National	America	White
Agriculturist	Standard	American	Davis
A Happy Home Maker	A. G. Mason Mfg. Co.	American	National
Aimsee	Davis	American Beauty	Davis
Ajax	Davis	American Empire	National
Akron	Free	American Junior	Standard
Alagame	Davis	American Lady	National
Alamo City	National	American National	National
Alaska	National	American, New	National
Albaugh Dover	National	American Queen	Goodrich
Albermarle	National	American Rotary	National
Alberta Peterson	Standard	American Standard	Standard

Sewing-Machine Name	Sewing-Machine Company	Sewing-Machine Name	Sewing-Machine Company
American Union	New Home	Arms	A. G. Mason Mfg. Co.
American Wonder	National	Armstrong	A. G. Mason Mfg. Co.
Americana	Standard	Arnold	A. G. Mason Mfg. Co.
Americo	National	Arnold Arrington	White
Amerikan	New Home	Arnold Overseam Sewing	
Amherest	Davis	Machine	Standard
Ami	Standard	Arrow	National
Amidon Bros.	A. G. Mason Mfg. Co.	Arrow	Standard
Amory How and Company	White	Art Superb	National
Amsler	White	Artison	Standard
Anchor	A. G. Mason Mfg. Co.	Ashland	National
Anchor	White	Asia	White
Anderson	Davis	Aston Special	A. G. Mason Mfg. Co.
Anderson	Goodrich	Astor, The	Standard
Anderson (A, B, C)	Davis	Atchison	National
Andrews	A. G. Mason Mfg. Co.	Athens	A. G. Mason Mfg. Co.
Andrus Rotary	White	Atkins, The	A. G. Mason Mfg. Co.
Angelus	Davis	Atlanta	National
Anita	National	Atlantic	Davis
Annabelle	National	Atlantic	Standard
Anthracite	Standard	Atlantis	New Home
Antharcits	White	Atlas	National
Antic	Goodrich	At-Will	Davis
Antitrust	National	Aubrey	White
Antonio Castillero	Standard	Auburn	Free
Apartment	Free	Auburndale	Standard
Apax	Davis	Auerbach	National
Apeth	National	August Junior	National
Apex	National	Aurora	Free
Appleton	New Home	Austin Special	National
Appolo	Free	Austral	Standard
Appolo	Standard	Austral, New	A. G. Mason Mfg. Co.
Apostulu Baptist	Standard	Austral Special	National
Arbest	New Home	Auto, The	Free
Arbutus	A. G. Mason Mfg. Co.	Autocrat	National
Arcade	National	Automatic Rotary	National
Arcade	Standard	Automatic, The	Standard
Arcadia	Standard	Autostitch	National
Arey	A. G. Mason Mfg. Co.	Auto Trust	National
Arganaut	Davis	Avena	A. G. Mason Mfg. Co.
Argote Lopez Co.	A. G. Mason Mfg. Co.	Avenue	White
Argyle	Davis	Avery	Free
Arkadelphia	A. G. Mason Mfg. Co.	Avia	A. G. Mason Mfg. Co.
Arlington	Davis	Aviator	Free
Arlington Anderson	Davis	Avon	National
Arlington Cash	Davis	Avona	New Home
Arlington Gem	National		
Arlington Gem	Standard	Baby Grand	Standard
Arlington Jewell	Davis	Baca and Bohac	Standard
Arlington Queen	National	Bacha	Standard
Armbrust	A. G. Mason Mfg. Co.	Backer	Free
Armistead	White	Bacon	A. G. Mason Mfg. Co.
Armour D	Standard	Baddley	White

Sewing-Machine Name	Sewing-Machine Company	Sewing-Machine Name	Sewing-Machine Company
Badger	Badger, Fond du Lac, Wisconsin (1907)	Batrow	White
		Batte Furniture Co.	A. G. Mason Mfg. Co.
Badt	National	Battermans	Free
Bailey	Goodrich	Battermans Special	Free
Baileys	White	Batiste Toledo	A. G. Mason Mfg. Co.
Bain	Standard	Baucan, C.	White
Bain	White	Bauerman	Davis
Baker	Goodrich	Bauermeister	Davis
Baldwin	Goodrich	Baumgarten	White
Baldwin	Davis	Baumgartner	White
Baldwin, New	Davis	Bauman	A. G. Mason Mfg. Co.
Bale, N. Y.	A. G. Mason Mfg. Co.	Baumert	White
Ball	White	Bauzon, Chas.	A. G. Mason Mfg. Co.
Ball Bearing	White	Baxendale	National
Ballenger	Goodrich	Baxter	Standard
Ballentinees	Standard	Bayless	White
Balliet	National	Bayounes	A. G. Mason Mfg. Co.
Balinoral	National	Bay State	Davis
Baltimore	National	Bazar	National
Baltimore Special	Davis	Beach	Davis
Bamberger	Davis	Beach	White
Bamberger	White	Beacon	National
Bancon	A. G. Mason Mfg. Co.	Beacon Imperial	A. G. Mason Mfg. Co.
Banfield	Goodrich	Beacor D	Davis
Banks	White	Beade, The	Standard
Banncr	Davis	Beagle Hole	Standard
Banner	National	Beahm, Geo.	Standard
Bannerman	National	Bean	Standard
Banques	Goodrich	Bean Special	Free
Banta	National	Beance	National
Baptiste	Goodrich	Bear	Davis
Barbosa Coraceivich	National	Beardsley	White
Bardwell	White	Beasley Special	A. G. Mason Mfg. Co.
Barfield	White	Beathe	Standard
Barlow	A. G. Mason Mfg. Co.	Beattie Special	A. G. Mason Mfg. Co.
Barnards	White	Beaumont	White
Barnes	A. G. Mason Mfg. Co.	Beauty	Free
Barnett	Davis	Beaver	National
Barnett	White	Bebauer	Standard
Barringer	White	Beck McK. Co.	White
Barron	White	Bedford	National
Barthalomew	White	Beegle	Davis
Barthalomew	A. G. Mason Mfg. Co.	Bee Hive	A. G. Mason Mfg. Co.
Bartigo Farmer	National	Bee Hive	Standard
Bartlett	Davis	Beesons	White
Bartlett	White	Beham	Standard
Bartlett Rotary	National	B. E. Harmon's Rotary	White
Bass	Free	Behlmers	Standard
Bastedor O Damode	National	Behrenden	National
Baston, D.Y., Cia	National	Behrens	A. G. Mason Mfg. Co.
Bates	A. G. Mason Mfg. Co.	Beigle	Davis
Batesville	National	Beirne	Goodrich
Batron	White	Belair	National

Sewing-Machine Name	Sewing-Machine Company	Sewing-Machine Name	Sewing-Machine Company
Belanger	A. G. Mason Mfg. Co.	Billedud	White
Belanger Special	Standard	Bill Morris Co.	Standard
Bel Isle	National	Bingham	Davis
Bell	Goodrich	Binghamton	Standard
Bell Bergum	White	Bishop Rotary	White
Bell Queen	National	Bishop	A. G. Mason Mfg. Co.
Belle	National	Bishops Favorite	Standard
Bellvge	Davis	Bitkers	Free
Belmars	Standard	Bitkers Favorite	Free
Belmead	National	Bivens and Co.	A. G. Mason Mfg. Co.
Belmont	National	Bixby and Cox	Standard
Belmont Special	Free	Blackburn	Free
Belvidere	National	Black Diamond	Standard
Belview	Goodrich	Blackey	A. G. Mason Mfg. Co.
Ben Hur	Davis	Blackhawk	National
Benjamin	Standard	Blacknell	Standard
Benjamin	National	Black Leader	New Home
Bennetts Best	A. G. Mason Mfg. Co.	Blacks Reliable	A. G. Mason Mfg. Co.
Bennita	National	Blacks Special	New Home
Bentons Favorite	A. G. Mason Mfg. Co.	Blade	Davis
Benty	National	Blade	Goodrich
Benway Special	Free	Blade, New	Goodrich
Berkle	National	Blahm	White
Berkley	National	Blank	A. G. Mason Mfg. Co.
Berkshire Rotary	National	Blanton	White
Berstrands	White	Blaunt	White
Berta	National	Blocks	National
Bertha	Standard	Blocko	A. G. Mason Mfg. Co.
Bertig Bros.	National	Blount and E	Standard
Berwick	Free	Blount Special	Standard
Bessie	Standard	Blossom Hardware Co.	A. G. Mason Mfg. Co.
Bessies	A. G. Mason Mfg. Co.	Blue Bell	Standard
Bessies Best	National	Blue Bird Rotary	Davis
Bession	A. G. Mason Mfg. Co.	Blue Diamond	Standard
Best, The	A. G. Mason Mfg. Co.	Blue Diamond	Goodrich
Bestor	Standard	Blue Field	Goodrich
Bethds	White	Blue Ribbon	Davis
Betsy Ross	National	Blue Ridge	National
Betz	National	Bluff City	National
Beuleys	White	Boans	National
Beulah, The	National	Boanus	National
Beverly	White	Boass	National
Biddle	Davis	Bob Laflin	A. G. Mason Mfg. Co.
Biddy	A. G. Mason Mfg. Co.	Bob Stanley	Goodrich
Bierchwals	A. G. Mason Mfg. Co.	Boehm	A. G. Mason Mfg. Co.
Bierchwale	White	Boetler Special	National
Biffor	Goodrich	Boettcher	White
Big Four	White	Boggs and Buhl	Standard
Big Sandy	A. G. Mason Mfg. Co.	Bohn	Davis
Big Store	A. G. Mason Mfg. Co.	Bollo Hardware Co.	A. G. Mason Mfg. Co.
Biggs	A. G. Mason Mfg. Co.	Bolton	Goodrich
Biggs and Hanslip	Standard	Bon March	Standard
Billeand	A. G. Mason Mfg. Co.	Bon, The	Davis

Sewing-Machine Name	Sewing-Machine Company	Sewing-Machine Name	Sewing-Machine Company
Bonanza	A. G. Mason Mfg. Co.	Brenners Rotary	National
Bonaventure	National	Brenton and H.	A. G. Mason Mfg. Co.
Bond, T., Co.	White	Brewer A, B, C	Free
Bond, T., Hardware Co.	New Home	Brewer AE-BE-DE	National
Bond	White	Brewer Electric	National
Bonia	New Home	Brewer Electric R	National
Bon-I-Look	Standard	Brewer O-Keh	New Home
Bonita	National	Brewer Rotary	White
Bonley	A. G. Mason Mfg. Co.	Brewer Rotary E	National
Bonner	Standard	Brickel	White
Bonota	National	Bride	National
Boody Bros.	A. G. Mason Mfg. Co.	Briggs	White
Booker	White	Bright Star	National
Boones Leader	A. G. Mason Mfg. Co.	Brigton	Davis
Boose and Buhl	Standard	Brin	A. G. Mason Mfg. Co.
Booster	Goodrich	Brisco	National
Booth	White	Bristol	Goodrich
Booty	A. G. Mason Mfg. Co.	Britian	Davis
Borden City	National	Broadway	Standard
Borg	Standard	Broadway	White
Borg Special	National	Broatbryay Special	National
Boss, The	National	Broken Arrow	Standard
Boston	Davis	Brock	White
Boston A	Free	Brodbents	White
Boston Grand	National	Broes	White
Boston Rotary	Standard	Brooker	White
Boston S	White	Brookly Edison Special	A. G. Mason Mfg. Co.
Boston Store Rotary	Standard	Brooklyn Queen	A. G. Mason Mfg. Co.
Bouaters	A. G. Mason Mfg. Co.	Brooks	White
Bouers	White	Brooks Light Running	Standard
Bouquett	Free	Brooks Special	Free
Bourland	White	Brookshire	Goodrich
Boutell	National	Browholm	White
Bowen	White	Brown	A. G. Mason Mfg. Co.
Bowerman	A. G. Mason Mfg. Co.	Brown Special	Standard
Box Walter	A. G. Mason Mfg. Co.	Brownie	A. G. Mason Mfg. Co.
Boyless	White	Browns	Davis
Bracht	White	Browns	Standard
Bracht Bros.	A. G. Mason Mfg. Co.	Bruce	A. G. Mason Mfg. Co.
Bradbury	White	Bruce	Free
Bradford	Foley & Williams Mfg. Co.	Bruce	Standard
		Bruce Rotary	White
Bradford	Goodrich	Brumhall	Davis
Bradley	A. G. Mason Mfg. Co.	Brunswick	National
Brady Bros.	A. G. Mason Mfg. Co.	Bry-Special	Davis
Bramer	A. G. Mason Mfg. Co.	Bryan	A. G. Mason Mfg. Co.
Bramlett	A. G. Mason Mfg. Co.	Bryant	White
Bramor	National	Brys	Goodrich
Brant	Davis	Brucker	A. G. Mason Mfg. Co.
Brasselton	A. G. Mason Mfg. Co.	Bucker	A. G. Mason Mfg. Co.
Bread Winner	National	Buckey	Davis
Breedys Best	National	Buckeye	Davis
Breham	Standard	Buckeye	Standard

Sewing-Machine Name	Sewing-Machine Company	Sewing-Machine Name	Sewing-Machine Company
Buckley Bros.	A. G. Mason Mfg. Co.	Cactus	Goodrich
Budd	White	Cadilac I	National
Buena	Davis	Cadilac III	Standard
Buetners	National	Caer	Davis
Buffalo	Free	Caines Special	Standard
Buffalo	Goodrich	Calandric	Standard
Buffalo Queen	National	Caldwell and L	A. G. Mason Mfg. Co.
Buffalo Special	Standard	Calfat	National
Buhl	National	California	Goodrich
Bullock	White	Callaway	Standard
Bunch	White	Calumet	Free
Bungalow	National	Calyx	Standard
Bunker Hill	A. G. Mason Mfg. Co.	Calyx	White
Bunnell Special	Goodrich	Calyx Rotary	A. G. Mason Mfg. Co.
Bur Oak	National	Cal-YX Rotary	White
Burchwale	National	Cambria	Davis
Burdick 1-A	National	Cambridge	New Home
Burdick 2-B	Davis	Cameo	Goodrich
Burdick 3-C	Household	Cameran	Goodrich
Burdick 4-D	Free	Cammilto	Free
Burg	Davis	Campania	Standard
Burgess Special	Free	Campbell	Standard
Burgess, The	A. G. Mason Mfg. Co.	Campbells Gem	A. G. Mason Mfg. Co.
Burhart	White	Cams Special	Standard
Buris and K.	A. G. Mason Mfg. Co.	Canadian	National
Burks	A. G. Mason Mfg. Co.	Canadian Empire, The	National
Burks	White	Cannons	A. G. Mason Mfg. Co.
Burlington	National	Canton	A. G. Mason Mfg. Co.
Burnett	Goodrich	Capac Star	A. G. Mason Mfg. Co.
Burnetts Choice	National	Cape Fear	Free
Burns	A. G. Mason Mfg. Co.	Capital	National
Bur Oak	National	Capitol	A. G. Mason Mfg. Co.
Burrleson	White	Capitol	Standard
Burrous Gem	A. G. Mason Mfg. Co.	Capitola	Free
Burtis Bros.	White	Car	National
Burton	National	Caraflor	White
Burton & Co.	White	Caraway	Standard
Busbee	White	Cardoza	Davis
Bush	White	Cardozas	Goodrich
Busha Special	A. G. Mason Mfg. Co.	Carle	White
Buster Brown	A. G. Mason Mfg. Co.	Carlin & F	Davis
Busy Bee	National	Carlisle	A. G. Mason Mfg. Co.
Busy Woman	National	Carlock and R.	A. G. Mason Mfg. Co.
Butler	Free	Carmelita	National
Butler Special	Free	Carmicheal	National
Buttruff	A. G. Mason Mfg. Co.	Carnahan	Standard
Byron	Free	Carnahan	A. G. Mason Mfg. Co.
Byrson and G.	National	Carney, The	A. G. Mason Mfg. Co.
Bywater Special	A. G. Mason Mfg. Co.	Carnival	Davis
		Carola	National
Cabinet	National	Carolina	National
Cables	New Home	Carolyn	White
Caborn	Goodrich	Carpenter	White

Sewing-Machine Name	Sewing-Machine Company	Sewing-Machine Name	Sewing-Machine Company
Carr Bros.	White	Chandler D. W.	Standard
Carr Special	National	Chaney	A. G. Mason Mfg. Co.
Carrara	National	Chant and C.	National
Carren	National	Chappel, The	Free
Carroll	Free	Chapman	National
Carson Special	A. G. Mason Mfg. Co.	Charlotte	National
Carswell	Davis	Charm	A. G. Mason Mfg. Co.
Carter	Goodrich	Charmer	National
Carthage Special	A. G. Mason Mfg. Co.	Charmount, La.	A. G. Mason Mfg. Co.
Carthagenian	Standard	Charter Oak	Davis
Carycon	A. G. Mason Mfg. Co.	Charter Oak	National
Casa Bicalko	A. G. Mason Mfg. Co.	Chase City	National
Casa Eperanca	A. G. Mason Mfg. Co.	Chatham	National
Casa Florenzeno	A. G. Mason Mfg. Co.	Chattanooga	Davis
Cascade	Davis	Chautauqua	New Home
Case, The	National	Cheerful Moment	National
Cases	National	Cheiftain	Davis
Cases Leader	National	Chelsea	Davis
Cash Department Store	Goodrich	Cheming	Davis
Casino Special	Standard	Cherokee	A. G. Mason Mfg. Co.
Casses Number I	National	Cherokee	Free
Cates	White	Chesapeake	National
Cats	White	Chester	White
Causeway	Davis	Chicago	National
Cayce Publishing Co.	Free	Chicago	Goodrich
Cecil	A. G. Mason Mfg. Co.	Chicf Wapello	Davis
Celeste	National	Childress	A. G. Mason Mfg. Co.
Celeste	A. G. Mason Mfg. Co.	Childs Special	National
Centenial	A. G. Mason Mfg. Co.	Chiles	National
Centenial	Standard	Chillico Hardware Co.	A. G. Mason Mfg. Co.
Centerfield	Standard	Chillicothe	White
Center and P.	A. G. Mason Mfg. Co.	Chimanco	National
Central City	Free	China	National
Central Methodist	National	Chinic	Davis
Central Model	Goodrich	Chisms Leader	A. G. Mason Mfg. Co.
Centric	Standard	Chan	White
Centura	Davis	Christian Courier	National
Century	Davis	Christie	National
Century Grand	National	Christine	A. G. Mason Mfg. Co.
Century Mfg. Co.	Free	Chronicle, New	National
Century Mfg. Co.	National	Chronicle Premium	Goodrich
Century Rotary	White	Chubert	Standard
Ceres	National	Chum and B.	A. G. Mason Mfg. Co.
Chachere	Standard	Cisco	Davis
Challenge	National	Clarborn	Standard
Challenge	Davis	Clarlender	National
Chamber, A. K.	A. G. Mason Mfg. Co.	Clark	Davis
Chamber Bar	National	Clark, F. S.	A. G. Mason Mfg. Co.
Chamber Domestic	Goodrich	Clark, J. A.	Standard
Chamber, New	Free	Clarke Rotary	White
Chambert Rotary	White	Class	A. G. Mason Mfg. Co.
Champion	National	Classic	National
Champlain	Standard	Clauson and Wilson	A. G. Mason Mfg. Co.

Sewing-Machine Name	Sewing-Machine Company	Sewing-Machine Name	Sewing-Machine Company
Clays Special	National	Commercial	Standard
Clayton	National	Commercial Electric	New Home
Clemons Bros.	A. G. Mason Mfg. Co.	Common Sense	A. G. Mason Mfg. Co.
Cleveland	Davis	Common Wealth	Davis
Cleveland, The	Standard	Common Wealth	Free
Cleveland Gem	A. G. Mason Mfg. Co.	Companion City	National
Cleveland Peerless	A. G. Mason Mfg. Co.	Competition	National
Clifford	National	Comstock Electric	New Home
Cliftoa	National	Comus	National
Cliftoa	Free	Concha	A. G. Mason Mfg. Co.
Climax	New Home	Concordia	National
Clindison Special	National	Concra	White
Clipper	New Home	Conley	Standard
Clouer	A. G. Mason Mfg. Co.	Conley	White
Clover	Davis	Conlon	White
Club, The	Standard	Connells Leader	National
Cobbs	White	Connolly	White
Cobbs Special	A. G. Mason Mfg. Co.	Conover	Davis
Cockran	National	Conover, B.	Davis
Codgel Special	A. G. Mason Mfg. Co.	Conover, The	A. G. Mason Mfg. Co.
Coffee, J. W.	New Home	Conquest	National
Coggins, B. F.	A. G. Mason Mfg. Co.	Conquerer	National
Coin, The	Free	Conshatter	Standard
Cohns	National	Constitution	Goodrich
Cole	Standard	Constitution	National
Cole Special	A. G. Mason Mfg. Co.	Constitution Atlanta	Goodrich
Coleberg	Free	Continental	National
Co-Lee	A. G. Mason Mfg. Co.	Continental Rotary	National
Collers	A. G. Mason Mfg. Co.	Convery	White
Collins	National	Conway & Jordan	A. G. Mason Mfg. Co.
Colonial	Davis	Coogans 20th Century	Standard
Colonial	New Home	Cooks	A. G. Mason Mfg. Co.
Colonial	Standard	Cooley, The	Standard
Colonial House	National	Coopers	A. G. Mason Mfg. Co.
Colonial Steinway	National	Co-operative	A. G. Mason Mfg. Co.
Colorado	Standard	Copeland	A. G. Mason Mfg. Co.
Colquit	A. G. Mason Mfg. Co.	Copeland	National
Columbia	Davis	Coquett	Standard
Columbia	New Home	Corasco	Goodrich
Columbia Record	National	Cordts	White
Columbian	National	Corinth	White
Columbine	Standard	Corley	Goodrich
Columbus	New Home	Corley, The	Standard
Columbus	Davis	Corymer	White
Columbus, The	A. G. Mason Mfg. Co.	Corn Belt	A. G. Mason Mfg. Co.
Columbus, R.	White	Cornelius	A. G. Mason Mfg. Co.
Columbus Special	National	Cornell	Standard
Comanche	White	Cornellys Leader	A. G. Mason Mfg. Co.
Combest	A. G. Mason Mfg. Co.	Corner, The	Standard
Comfort	A. G. Mason Mfg. Co.	Cornerstone	National
Commander	Free	Corola	Davis
Commercial	Davis	Coronell	Davis
Commercial	National	Coronet	Standard

Sewing-Machine Name	Sewing-Machine Company	Sewing-Machine Name	Sewing-Machine Company
Corrara	National	Cuba Libre	Standard
Corrolla	A. G. Mason Mfg. Co.	Cubana	Goodrich
Correct	Free	Culler, W. C.	A. G. Mason Mfg. Co.
Cosmo	New Home	Culliman	Standard
Cosper	White	Cullum, L. D.	A. G. Mason Mfg. Co.
Cosseboom	Standard	Cumberland	Davis
Cottage	New Home	Cumberland Special	Free
Cottage	National	Cummings Special	A. G. Mason Mfg. Co.
Cottage Queen	New Home	Currie	A. G. Mason Mfg. Co.
Cotton	Standard	Cutlers	Davis
Couilliard	National	Cuyahoga	A. G. Mason Mfg. Co.
Coulson	Davis	Cuyahoga	Standard
Counters	Davis	Czar	Goodrich
Countess	A. G. Mason Mfg. Co.		
Co-up	White	D'Amour	National
Courier	National	Daffodil, The	National
Courier Journal	National	Da Horn Sing	Davis
Couan	A. G. Mason Mfg. Co.	Dahlichs	A. G. Mason Mfg. Co.
Covington Co.	A. G. Mason Mfg. Co.	Dailey and Co.	National
Cowell & S.	White	Dailey, W. S.	Standard
Cox	White	Daisey	Standard
Coyne	Free	Daisey	Davis
Craig	White	Daisey	Goodrich
Craig Rotary	White	Daisey	A. G. Mason Mfg. Co.
Cramer	White	Dalay San Sords	White
Cranberry	Free	Dallas, New	A. G. Mason Mfg. Co.
Crass	White	Dallas-W-Co.	White
Craver & E	White	Dallier	White
Crawcord	White	Damascus	Standard
Crawford	Davis	Damascus	National
Crawford	Standard	Danake	National
Crawford	White	Dandremon	Goodrich
Crawleys	White	Dandy	Goodrich
Creed	White	Dandy	New Home
Creedmoor	Standard	Dandy, New	Standard
Cremo	White	Dan Fords	White
Cresent	Davis	Danforth	A. G. Mason Mfg. Co.
Cresent	Free	Dania	A. G. Mason Mfg. Co.
Cresent	Standard	Daniels	A G. Mason Mfg. Co.
Cresent	White	Daniels	Standard
Crippin	Household	Danskeren	Goodrich
Criterion	National	Danson	National
Croley	White	Darlington	National
Crook-Record	White	Darner Bros.	White
Crosby	White	Darrison	White
Cross	A. G. Mason Mfg. Co.	Daughtys	White
Crown	A. G. Mason Mfg. Co.	Duske's Pioneer	National
Crown Jewell	Free	Dauntless	National
Crown Rotary	White	Dauntless	New Home
Crump	White	Dauphine	National
Crusader, The	Standard	Davenport	National
Cruse, E.	A. G. Mason Mfg. Co.	Davidson	National
Crutchen, C. M.	Standard	Davies	Davis

Sewing-Machine Name	Sewing-Machine Company	Sewing-Machine Name	Sewing-Machine Company
Davis	Davis	De Sota	National
Davis, A. J.	White	Despatch	National
Davis, A. J. Son	A. G. Mason Mfg. Co.	Detonia	Davis
Davis, C. M.	White	Detroit	National
Davis Rotary C. D. 1. 2.	National	Devereaux	National
Dawson Special	Free	Deverlax	White
Dawley 3	Goodrich	Devon	National
Day's Special	National	Dewey	A. G. Mason Mfg. Co.
Dayton	Davis	Dewey, The	Standard
Daytonia	Davis	De-Will Co.	White
Dean Special	A. G. Mason Mfg. Co.	Dewitt	National
Dean, The	Standard	Dey's	White
Dearing	National	Diabolo	White
De Boren	National	Diamond	National
Decatur	National	Diamond D	Standard
Decatur, The	Standard	Diamond-Models	National
Decora Special	National	Diamond Royal	Free
Decorah Posten	National	Diana	National
Defender	A. G. Mason Mfg. Co.	Dickson	A. G. Mason Mfg. Co.
Defiance	Davis	Dietzel	A. G. Mason Mfg. Co.
Defiance	Standard	Dispatch	A. G. Mason Mfg. Co.
Defiance	National	Dittenger	A. G. Mason Mfg. Co.
De Kahl	A. G. Mason Mfg. Co.	Ditzell Imp.	National
Delamer Elect	National	Ditzer	National
Delanorton	Goodrich	Dixie	Davis
Delaplane	White	Dixie	Free
Delaware	Davis	Dixie	National
Delia	Standard	Dixie	Standard
Delight	Davis	Dixie Automatic	New Home
Delight	Free	Dixie Queen	National
Delight	National	Dixon	National
Delmar	National	Dobie Special	New Home
Delong	White	Dobbs	A. G. Mason Mfg. Co.
Delphia	Free	Dodge Rotary	National
Delta	Free	Dodds	A. G. Mason Mfg. Co.
Deluxe	National	Dodge and Watson	Standard
Demerest Dry Co.	White	Dodson and Rice	Standard
De Morley	National	Dome Rotary	National
Democrat	National	Domestic Buffalo	Domestic
Dempsters	National	Domestic, New	Goodrich
Den	National	Domestic, Old	Goodrich
Den Dansko	National	Domestic Rotary	Domestic
Denholm	Standard	Domestic Science	National
Dennis	A. G. Mason Mfg. Co.	Dominion	National
Dennis Special	Davis	Dominion, The	Free
Dennison	National	Donaldson's Rotary	White
Dentz	A. G. Mason Mfg. Co.	Donalds Special	A. G. Mason Mfg. Co.
Denver	Davis	Donohue, T.	White
Denver	National	Dorcas	National
Denver	Standard	Dorcas	A. G. Mason Mfg. Co.
Departure	National	Dormer	A. G. Mason Mfg. Co.
Dependable	Standard	Dorothea	A. G. Mason Mfg. Co.
De Sola	National	Dorroh-K Co.	White

Sewing-Machine Name	Sewing-Machine Company	Sewing-Machine Name	Sewing-Machine Company
Dorsey	A. G. Mason Mfg. Co.	Eatonia	National
Dosenuna	Standard	Eberhart and B.	White
Dotheron	Standard	Eberlain Flyer	Standard
Doty	Free	Eblem	White
Douglas	Standard	Eclipse	Davis
Douglas	A. G. Mason Mfg. Co.	Eclipse	National
Downes	Standard	Eclipse	New Home
Downes	A. G. Mason Mfg. Co.	Economist	White
Drake	National	Economy	National
Drennen	A. G. Mason Mfg. Co.	Economy	Domestic
Dressmaker	National	Economy	New Home
Dreus and Crens	Standard	Economy Rotary	Domestic
DrisKell	A. G. Mason Mfg. Co.	Eddy	White
Droyer	National	Edgemere	National
Druid	National	Edgmore	Goodrich
Dryden	White	Edgwater	Standard
Duchess	National	Edgwood	Standard
Duchess	A. G. Mason Mfg. Co.	Edison A, B, C, D, E, F, G	Goodrich
Duhon	White	Edison, The	Standard
Duker	National	Edison Rotary	Standard
Dulitz C	Standard	Edison Rotary	White
Du Lux	Goodrich	Edwards	Davis
Dumore	National	Edwards	White
Duncan	A. G. Mason Mfg. Co.	Edwards, The	Free
Dundee	Standard	Edwards & Bradford	National
Dunham	Davis	Efficient	Davis
Dunklins	A. G. Mason Mfg. Co.	Ehlert	White
Dunlaps Gem	A. G. Mason Mfg. Co.	Elbeco	Davis
Duplex	National	Elberta	National
Duplex Rotary	White	Elbin	Free
Du Quesne	Free	El-Carreo	Standard
Durant	A. G. Mason Mfg. Co.	El-Cente Merco	A. G. Mason Mfg. Co.
Durgan	White	El-Center	A. G. Mason Mfg. Co.
Durham	Household	Elcock	White
Durrell	A. G. Mason Mfg. Co.	El-Diamante	Goodrich
Durrett	White	Eldorado	National
Duskell	A. G. Mason Mfg. Co.	Eldoro	Standard
Duske's Pioneer	National	Eldredge	National
Duval	Goodrich	Eldredge Rotary	National
Duyer	Goodrich	Eldredge 2-spool	National
Dutchess	National	Eldson	Free
Dyson	White	Eldson Imperial	Free
		Eld, The	National
Eagle	National	Electric	Davis
Eagle	Standard	Electric, The	New Home
Eagle Rotary	White	Electric City	National
Eames	Goodrich	Electric Supply Company	New Home
Earl	National	El-Galto	National
Earlham	National	El-Genio	National
Easley and D.	White	Elgin	Davis
Eastaluchia	Standard	Elgin, The	Free
Eastern	National	El-Globo	National
Easy Rotary	White	El-Hagar	A. G. Mason Mfg. Co.

Sewing-Machine Name	Sewing-Machine Company	Sewing-Machine Name	Sewing-Machine Company
Elherta	National	Englewood	National
Eliasburg	White	English	White
Elia	National	Enterprise	National
Elite	Free	Enterprise Rotary	White
Elite	National	Envoy	National
Elite Rotary	White	Eopanola	White
Elizabeth	A. G. Mason Mfg. Co.	Epworth	New Home
Elkhart	White	Era	National
Elkins	A. G. Mason Mfg. Co.	Era	Goodrich
Ellenhurst	National	Erie	National
Ellens Tar	A. G. Mason Mfg. Co.	Erie Special	Standard
Elliot	Standard	Ermine	National
Elliot	White	Erwin	White
Elliptic	National	Essex	Davis
Ellis	Goodrich	Essex	A. G. Mason Mfg. Co.
Ellis	Standard	Estell	A. G. Mason Mfg. Co.
Ellis Store Co.	White	Estey, New	National
El-Louvre	A. G. Mason Mfg. Co.	Euclid	Standard
Elly	National	Euclid	White
Elma	National	Eudora	National
Elmira	National	Eunice Hdw.	White
Elmo	Free	Eurcka	Davis
Elmore	Standard	Eureka	New Home
Elmore, New	National	Evangelistern	Goodrich
El-Niagra	National	Evans	National
El-Nico	A. G. Mason Mfg. Co.	Everett, The	National
El-Nouvre	A. G. Mason Mfg. Co.	Evergood A, B, C, D, E.	National
El-Nueva Coique	Standard	Everhart	White
El-Parrae	A. G. Mason Mfg. Co.	Ever Ready	Foley & Williams
Elpico	White	Ever Ready	Goodrich
El-Ralampage	Standard	Everybodys	Standard
Elro	National	Everybodys	White
Elsie	Standard	Everyday Life	Free
El-Singer Fatma	Standard	Excella	White
El-Sol Telkart	National	Excellent	National
Elva	National	Excellent	Standard
Elwess	National	Excello	Free
Ely	Unknown	Excells	Free
Emblem	New Home	Excelsior	Davis
Emerson	National	Excelsior	National
Emilia	White	Excelsior	New Home
Emma	National	Excelsior	Standard
Emmons	National	Expert	New Home
Emperor	White	Expert	Standard
Empire	Davis		
Empire	New Home	Fair	Davis
Empire State	A. G. Mason Mfg. Co.	Fairbanks	Davis
Emporium	Goodrich	Fairbanks	White
Empress	Davis	Fairlay	Standard
Empress	Free	Fair, The	A. G. Mason Mfg. Co.
Endora	National	Fairy	National
Enfield	A. G. Mason Mfg. Co.	Fairview	National
England	National	Fait, H.	National

Sewing-Machine Name	Sewing-Machine Company	Sewing-Machine Name	Sewing-Machine Company
Faithful Friend	Household	Fewells	National
Falcon	New Home	Fidelity	A. G. Mason Mfg. Co.
Falcous Gem	National	Fielden	White
Fallow	White	Fields	A. G. Mason Mfg. Co.
Falls City	Free	Fields Rotary	A. G. Mason Mfg. Co.
Falls City	Standard	Fienstein	National
Fambro	White	Fife	Goodrich
Fameron	National	Fikes	White
Family Favorite	New Home	Films Favorite	Goodrich
Family Friend	National	Fingers	White
Family Grand	National	Finglestein	Free
Family Jewel	National	Finkestein	Unknown
Family, New	A. G. Mason Mfg. Co.	Finkle and Lyon	A. G. Mason Mfg. Co.
Family Pride	National	Finklesstein	White
Family Queen	Standard	Finney	White
Family Rotary	White	Fireside	Davis
Famous	Davis	Fireside	New Home
Famous	National	Firk Special	A. G. Mason Mfg. Co.
Famous	Standard	Fischer	New Home
Famous Bar	White	Fischer	Standard
Farm & Ranch	National	Fischer	White
Farmers Alliance	National	Fish	White
Farmers Bride	A. G. Mason Mfg. Co.	Fitche Special	Standard
Farmers Friend	A. G. Mason Mfg. Co.	Fitzgerald	Standard
Farmer's Guide	National	Fitzgerald	A. G. Mason Mfg. Co.
Farmers Review	Davis	Flag	Standard
Farmers Rotary	A. G. Mason Mfg. Co.	Flanigan	Davis
Farmerson	Davis	Flanniggin Favorite	A. G. Mason Mfg. Co.
The Farmer's Superior Rotary	White	Flaum	White
Farmers Supply Co.	A. G. Mason Mfg. Co.	Fletwood	A. G. Mason Mfg. Co.
Farmers Union	National	Fletcher	National
Farmers Voice	Standard	Flint	A. G. Mason Mfg. Co.
Farro Special	A. G. Mason Mfg. Co.	Florals	National
Fashion	Davis	Flor De Maio	A. G. Mason Mfg. Co.
Fashion	Free	Flor De Dora	National
Fashion	Standard	Florence	A. G. Mason Mfg. Co.
Fashion, New	A. G. Mason Mfg. Co.	Florence Rotary	White
Faultless	Free	Florence Special	A. G. Mason Mfg. Co.
Faultless	National	Florence Vib.	A. G. Mason Mfg. Co.
Faultless	Standard	Florida	National
Faust	White	Flyer, The	Standard
Favorite	Davis	F. M. Rotary	White
Favorite	New Home	Foerste	Standard
Favorite	Standard	Footway, The	National
Favorite	White	Forbes Best	A. G. Mason Mfg. Co.
Fawl-Kas	White	Forbes S. G.	Standard
Fayette	National	Forbis	White
Feather	Household	Ford	National
Ferde De Siecle	Goodrich	Forest City	National
Fergeson	A. G. Mason Mfg. Co.	Forest City	Standard
Ferrell	White	Forester	White
Ferris	Standard	Forlows	White
Fessinger	A. G. Mason Mfg. Co.	Formon, D. H.	A. G. Mason Mfg. Co.

Sewing-Machine Name	Sewing-Machine Company	Sewing-Machine Name	Sewing-Machine Company
Forrester R.	White	Gainsborough	National
Forsythes	White	Galeusky	White
Fort Henry	National	Galicy Polski	Standard
Fort Worth	Davis	Galloway	National
Fort Worth Record	National	Galveston, New	National
Forward	A. G. Mason Mfg. Co.	Gamoriel	National
Foster	Goodrich	Ganda	National
Fountain	Davis	Gannon	National
Four Hundred, The	Standard	Garden City	Davis
Fowles, J. G.	A. G. Mason Mfg. Co.	Garden City	National
Foy and Gibson	White	Gardener Queen	A. G. Mason Mfg. Co.
Fram A	National	Gariey	A. G. Mason Mfg. Co.
Francis	A. G. Mason Mfg. Co.	Garland	Davis
Francis Wallard	National	Garland	New Home
Franenthal	White	Garland	Standard
Franklin	National	Garlicks	Standard
Franklin	White	Garlington	White
Franklin Bros.	White	Garrens	A. G. Mason Mfg. Co.
Franklin Hand A	Davis	Garry	White
Franklin, The	National	Garuer	White
Franks	Davis	Garvin	A. G. Mason Mfg. Co.
Free	Free	Carvis	A. G. Mason Mfg. Co.
Free Lock Stitch	Free	Gary	A. G. Mason Mfg. Co.
Free Press	Standard	Gastonia	A. G. Mason Mfg. Co.
Free Silver	Standard	Gates	Unknown
Free Trade	New Home	Gates City	Free
French, H	National	Gates City Turn. Co.	Free
French and Bassett	Free	Gaucha	National
Fricke	White	Gauger	National
Frictionless	Davis	Gayiety	National
Friday	National	Gaylor	National
Friend	White	Gayosa	National
Frisbis	White	Gazell	Goodrich
Frontier	Free	Geitz Rotary	White
Frozen Dog	Goodrich	Gellers	Davis
Fulga Brothers	National	Gem	Davis
Full Dinner Pail	National	Gem	Free
Fuller	National	Gem City	National
Fullmer	A. G. Mason Mfg. Co.	Genesee	National
Full Wide	A. G. Mason Mfg. Co.	Genesee	Standard
Fulner	White	Geneva	National
Fulton	Davis	Georgia	Davis
Fulton	National	Georgia Hardware Co.	A. G. Mason Mfg. Co.
Furgeson	White	Gerald	A. G. Mason Mfg. Co.
Furtick	Standard	Gerekes, J. S.	Free
Furney	A. G. Mason Mfg. Co.	Germeths Favorite	Standard
Furs	Standard	Germania	National
Futz	White	Germania	Goodrich
		Gertrude	Standard
Gable	A. G. Mason Mfg. Co.	Gholston	A. G. Mason Mfg. Co.
Gaddis	A. G. Mason Mfg. Co.	Gibbon	National
Gail	Davis	Gibson	White
Gaines	White	Gignilliant	Standard

Sewing-Machine Name	Sewing-Machine Company	Sewing-Machine Name	Sewing-Machine Company
Gilbert Queen	A. G. Mason Mfg. Co.	Goodes	Standard
Gillespic and M.	Standard	Good Luck	National
Gillespic, New	National	Goodness	A. G. Mason Mfg. Co.
Gills	White	Goodman	A. G. Mason Mfg. Co.
Gilmore, H. C., A, B	Free	Goodnuf	A. G. Mason Mfg. Co.
Gimbel	Davis	Goodsich	Goodrich
Gimbel	A. G. Mason Mfg. Co.	Goodson	A. G. Mason Mfg. Co.
Gimbel Bros.	Standard	Goodyear	Goodrich
Gimbel Rotary	White	Gordon Favorite	Standard
Gimbel Special	National	Gores Best	A. G. Mason Mfg. Co.
Ginger Loderna	Standard	Gorman	A. G. Mason Mfg. Co.
Ginsburg	Davis	Goshen	National
Girard	Free	Gosnells	A. G. Mason Mfg. Co.
Gladiator	Davis	Gospel Adv.	National
Gladiator	A. G. Mason Mfg. Co.	Gospel News	Standard
Glasgow	Goodrich	Goss and Wake	Standard
Glass Block	Goodrich	Gossett	Standard
Glasser	National	Gossett Special	A. G. Mason Mfg. Co.
Glen Miller	Davis	Gourlay	Standard
Glenn	A. G. Mason Mfg. Co.	Gourlay, New	A. G. Mason Mfg. Co.
Glenn Rotary	White	Gourley	Davis
Glen Oak	National	Govenor	New Home
Glenwood	A. G. Mason Mfg. Co.	Gracie	Standard
Globe	National	Graciosa	National
Globe	Davis	Grafton	National
Globela	National	Gragard	A. G. Mason Mfg. Co.
Glock Special	Standard	Graham	Goodrich
Glod Flash	White	Granada Stores	National
Gloria	Davis	Grand	Davis
Gloria	White	Grand	Goodrich
Gloria Rotary	White	Grand	National
Glorioso	A. G. Mason Mfg. Co.	Grand Bazaar	A. G. Mason Mfg. Co.
Goerke	White	Grand Bulletin	National
Goerke Rotary	White	Grand Bulletin	Goodrich
Goldberg	A. G. Mason Mfg. Co.	Grand Leader	Goodrich
Goldberg's Rotary	White	Grand Leader	Standard
Goldbound	Standard	Grand Pacific	National
Gold City	National	Grand Rotary	National
Gold Dust	National	Grand Rotary	Standard
Gold Eagle	Goodrich	Grand Union	Davis
Gold Hibbard	Davis	Grand Union	National
Goldmans	White	Grand Union	Standard
Gold Medal	National	Grand Union	A. G. Mason Mfg. Co.
Gold Medal Rotary	National	Grange	National
Gold Mohr	New Home	Granger	White
Goldsmith	Free	Grange Store	White
Golden Eagle	National	Granite State	Standard
Golden Giant	A. G. Mason Mfg. Co.	Granore	Goodrich
Goldenrod	Goodrich	Grant, The	Davis
Golden Rule	Davis	Grant	Standard
Golden Star	A. G. Mason Mfg. Co.	Grant's Gem	A. G. Mason Mfg. Co.
Golden West	National	Grass	A. G. Mason Mfg. Co.
Gooch	White	Gray	A. G. Mason Mfg. Co.

Sewing-Machine Name	Sewing-Machine Company	Sewing-Machine Name	Sewing-Machine Company
Grays	White	Halco Rotary	White
Grayums	White	Halcyon	Davis
Great Northern	Goodrich	Hales Special	Davis
Great Western	Free	Hale, The	National
Greeacre	Standard	Haleys	White
Greeg	A. G. Mason Mfg. Co.	Hall	A. G. Mason Mfg. Co.
Greeman	Unknown	Halley	Standard
Green	White	Halls Favorite	A. G. Mason Mfg. Co.
Greenbuck	Davis	Halls Special	Standard
Greenwood	National	Halls, The	National
Gregg	White	Halsey Rotary	White
Grempezuski	A. G. Mason Mfg. Co.	Hambro-wall	White
Grenada	A. G. Mason Mfg. Co.	Hamilton	Davis
Gretchen	A. G. Mason Mfg. Co.	Hamilton	Goodrich
Gresco-F Co.	White	Hamilton	Standard
Greyhound	New Home	Hamilton	White
Griffin	National	Hammond	A. G. Mason Mfg. Co.
Griffin and S.	White	Hampton	Standard
Griffith	National	Hampton, C. L.	A. G. Mason Mfg. Co.
Gross Special	A. G. Mason Mfg. Co.	Hancock	A. G. Mason Mfg. Co.
Gruene	White	Handy	National
Grunder	White	Handy	Standard
Gualtney-U Co.	White	Handy Sewer	National
Guanoco	White	Hannie	Standard
Guarantee	A. G. Mason Mfg. Co.	Hanson	Standard
Guard	Free	Hanson Choice	Free
Guernsey	Standard	Hansons	White
Guggenheimer	A. G. Mason Mfg. Co.	Hantic Morn	Standard
Guide	Free	Hapgood	Goodrich
Guilford	National	HA. Philadelphia	Davis
Guillett	White	HA Philadelphia Singer	Standard
Guiterrez and N.	White	Happy Hearts	Goodrich
Gulf City	Free	Happy Hearts	National
Gulf Queen	Davis	Happy Home	New Home
Gunnell	A. G. Mason Mfg. Co.	Harbours	White
Gunton	Standard	Harde and G.	White
Guthrie	White	Hardin	White
Guthrie	A. G. Mason Mfg. Co.	Hardine	Standard
Guyler	National	Hardis	A. G. Mason Mfg. Co.
Gwaltney, U.	A. G. Mason Mfg. Co.	Hardison	A. G. Mason Mfg. Co.
Gypsy	National	Hardy	Standard
		Hargets	Standard
Hackett	National	Hargis	White
Hacketo Beauty	National	Harkeys	A. G. Mason Mfg. Co.
Hadda	A. G. Mason Mfg. Co.	Harlam	Standard
Hagedorn	White	Harley-P.	White
Hager	Davis	Harmack	Davis
Haggard	White	Harman	White
Haines, O. S.	Goodrich	Harold	National
Haines, New	Goodrich	Harper	National
Hairs	White	Harris	Davis
Halcod	White	Harris	White
Halcomb	White	Harrison	National

Sewing-Machine Name	Sewing-Machine Company	Sewing-Machine Name	Sewing-Machine Company
Harrison	Standard	Heneger	A. G. Mason Mfg. Co.
Harry's Special	Free	Henrico	National
Hart	A. G. Mason Mfg. Co.	Henrietta	Davis
Hartberg	A. G. Mason Mfg. Co.	Henwood	Standard
Hartberger	White	Henry	White
Hartford, New	National	Hepperle	White
Hartman	National	Herald	Free
Hartman	Standard	Herbert	White
Hartman	White	Herbert, H	Standard
Hartmere	Standard	Hercules	National
Hartin	National	Herder	A. G. Mason Mfg. Co.
Harvard	Davis	Herman	A. G. Mason Mfg. Co.
Harvard	Free	Hermitage	National
Harvard	A. G. Mason Mfg. Co.	Herpolscheimer	National
Harvard	White	Herrick	A. G. Mason Mfg. Co.
Harverty A, B, C	Free	Herrington	A. G. Mason Mfg. Co.
Harvey, The	National	Hesplin	A. G. Mason Mfg. Co.
Harvester	White	Heurietta	White
Harwell	White	Hewitt	A. G. Mason Mfg. Co.
Haserot	Standard	Heydes Star	A. G. Mason Mfg. Co.
Haskett	A. G. Mason Mfg. Co.	Heydes, The	National
Haspador	Davis	Hgnemans	White
Haus	Davis	Hiawatha	National
Haverty Special A, B, C	Free	Hibbard	National
Hawisha	National	Hicks	A. G. Mason Mfg. Co.
Hawishe	National	Hickory	National
Hawkeye	Davis	Higginbothan	A. G. Mason Mfg. Co.
Hawkeye	Foley & Williams Mfg. Co.	Higgins	A. G. Mason Mfg. Co.
		High A Family	A. G. Mason Mfg. Co.
Hawkeye	Goodrich	High A Victoria	National
Hayden	Free	Hightburg	National
Hazard	A. G. Mason Mfg. Co.	Hillman	Standard
Hazel	National	Hills	Goodrich
Hazelhurst	White	Hines	Standard
Hde. Sol and Co.	National	Hines, The	National
Hearthstone	National	Hinson	A. G. Mason Mfg. Co.
Heathwood	Standard	Hinton	A. G. Mason Mfg. Co.
Heffner-D	White	Hitchcock	National
Hefron	Free	Hjort	White
Hegus, The	Free	Hobert	Davis
Heinz	White	Hochschild	Standard
Heinenway	White	Hodde	White
Heinrich	Davis	Hoddes	Standard
Helaingoford	White	Hoffman	Standard
Hellweil	A. G. Mason Mfg. Co.	Hoffman	White
Helping Hand	National	Hoffman Daisy	Free
Helpmuth	A. G. Mason Mfg. Co.	Holcomb	A. G. Mason Mfg. Co.
Helsingsfors	A. G. Mason Mfg. Co.	Holcyon	Davis
Hemingways	Standard	Holden	A. G. Mason Mfg. Co.
Henderson	Davis	Holds Flyer	Standard
Henderson	Standard	Holland	A. G. Mason Mfg. Co.
Henderson, B.	Davis	Holliday	Free
Henderson certified D	Davis	Holman	National

Sewing-Machine Name	Sewing-Machine Company	Sewing-Machine Name	Sewing-Machine Company
Holmes	White	Housemate	Standard
Holmes, New	Standard	Housewife	National
Homan	Davis	Houston	Standard
Homan	Standard	Houston Bros.	A. G. Mason Mfg. Co.
Home	Davis	Houston Post	Goodrich
Home	Standard	Howard	National
Home Circle	National	Howe	White
Home Comfort	National	Howe, Imp.	Standard
Home Favorite	National	Howe Modern	A. G. Mason Mfg. Co.
Home Journal	Free	Howe Rotary	Standard
Home Leader	National	Howe Special	Standard
Home Pride	National	Howell	White
Home Queen	Unknown	Howland	Davis
Home and Farm OS	National	Howry	White
Homer	National	Hoye	White
Homer Fritz	Standard	Hudgkins	New Home
Homers	National	Hudson	New Home
Homers Bride	A. G. Mason Mfg. Co.	Hudson	Standard
Homestead	Free	Hudson	White
Homestead Rotary	National	Hueffner	White
Homewood	Goodrich	Hughes	Standard
Honey Cutt	A. G. Mason Mfg. Co.	Huh Special	Davis
Honeymoon	Davis	Huletts	White
Hong Yuen Co.	Standard	Hull and Dutton	Standard
Honor Bright	Goodrich	Hulse	National
Hooker	A. G. Mason Mfg. Co.	Humberts	White
Hooper	Standard	Humble Pride	White
Hooper	White	Humeycutt	White
Hoosier Standard	Standard	Hummer	Davis
Hoover	A. G. Mason Mfg. Co.	Hummer	A. G. Mason Mfg. Co.
Hope, New	Davis	Humming Bird	Davis
Hopkins	Free	Hunter	White
Hordernia	White	Huntington	Free
Horner	White	Hues M. Guerro	National
		Huos	White
Horn, W. W.	A. G. Mason Mfg. Co.	Hurley's Rotary	White
Horton	White	Hurson	Davis
Ho-se-ma	New Home	Husband and Patrandry	A. G. Mason Mfg. Co.
Hospodor	National	Husvennen	National
Hot Kutter	National	Hustler	Free
Hotpoint Rotary	White	Hyde	White
Houghton	Standard	Hyde Park	Free
Hourely	White	Hyman	Standard
Housan	Standard		
Household	Household	Ida	Standard
Household Leader	Free	Idaho	National
Household Pride	Free	Idea	National
Household Queen	National	Ideal	Davis
Household Star	Free	Ideal	White
Household Supply Co.	National	Ideal, New	New Home
Housemaid	National	Ideal R.	White
Houseman	A. G. Mason Mfg. Co.	Iiola	White
Housemate	New Home	Imperial	Davis
Housemate Special	Standard		

Sewing-Machine Name	Sewing-Machine Company	Sewing-Machine Name	Sewing-Machine Company
Imperial	New Home	Jackson	Free
Imperial Beacon	A. G. Mason Mfg. Co.	Jackson	Standard
Imperial Chataqua	Free	Jackson Special	National
Imperial Columbian	National	James	Standard
Imperial Howe	Standard	James	White
Improved Howe	Standard	Jamie	Standard
Improved Model	Goodrich	Jamison	Standard
Improved New Leader	New Leader	Jarman	White
Improved Royal	Free	Jasper	Standard
Improved Southern	National	Jasnaik	A. G. Mason Mfg. Co.
Improved Spokand	A. G. Mason Mfg. Co.	Jathis	A. G. Mason Mfg. Co.
Improved Tropp	A. G. Mason Mfg. Co.	Jay Grand	White
Improved Vassar	Free	Jeans	A. G. Mason Mfg. Co.
Impsasttone	Davis	Jefferson	National
Index	Davis	Jeffrenoun	Standard
Index	National	Jeidel	A. G. Mason Mfg. Co.
Indiana	Davis	Jellico	White
Indiana Farmer	National	Jennings	New Home
Indianapolis Sentinel	Davis	Jerome	National
Industrious Hen	A. G. Mason Mfg. Co.	Jens, O. H.	National
Ingle	A. G. Mason Mfg. Co.	Jetters Gem	A. G. Mason Mfg. Co.
Ingle	White	Jettinghoff and B	A. G. Mason Mfg. Co.
Ingliss	Standard	Jewell	Davis
Inland Farmer 4	National	Jewell, The	New Home
Innes, The	Standard	Jewell	A. G. Mason Mfg. Co.
Insurgent	Davis	Jnusines	National
International	Davis	Joekel	A. G. Mason Mfg. Co.
International	New Home	Johnson	A. G. Mason Mfg. Co.
Inter Ocean	National	Johson	Davis
Invader	National	Joli	Standard
Invincible	Standard	Jones	Standard
Invocation	National	Jones	A. G. Mason Mfg. Co.
Icma	National	Jonesboro	Free
Iowa I	National	Jordan Brothers Rotary	White
Iowa II	Davis	Josefaal	Standard
Iowa IV	Standard	Joseph	A. G. Mason Mfg. Co.
Iowa Queen	National	Joske	National
Iracema	White	Joswick	White
Iracena	Standard	Journal	National
Iraquos	Standard	Joy	A. G. Mason Mfg. Co.
Irian Bros.	White	Jubilee	White
Ironton	National	Judas	Goodrich
Italia	White	Judel	A. G. Mason Mfg. Co.
Itta-Bena	White	June, H. A.	Goodrich
Ivanhoe	Standard	Juno	Davis
Iver Johnson	A. G. Mason Mfg. Co.	Juno	Goodrich
Iverness	Davis		
Iver, The	A. G. Mason Mfg. Co.	Kaczer	A. G. Mason Mfg. Co.
I.X.L.	National	Kaeppel	A. G. Mason Mfg. Co.
		Kahn's Pride	Standard
		Kalamazoo	Davis
Jackle	Goodrich	Kalamazoo	White
Jackson	Davis	Kamp	Standard

Sewing-Machine Name	Sewing-Machine Company	Sewing-Machine Name	Sewing-Machine Company
Kanawha	National	Kimble	A. G. Mason Mfg. Co.
Kane Special	A. G. Mason Mfg. Co.	Kimbrough	A. G. Mason Mfg. Co.
Kangaroo	Standard	Kinerys Star	A. G. Mason Mfg. Co.
Kangroo	White	King	National
Kanokla	Goodrich	King Bee	Free
Kansas	A. G. Mason Mfg. Co.	King Buffalo	New Domestic
Kansas City	Davis	King Cotton	Davis
Kansas Farmer	National	King Ford	National
Kantank	Free	King Land	National
Kantauk	Free	King, New	Davis
Karrs Special	A. G. Mason Mfg. Co.	King Perfection	National
Kasharek	White	King Star	A. G. Mason Mfg. Co.
Kasparek	Free	King Sterling	National
Kaufman A, B, C, D	National	King, The	Standard
Kavanaugh	Free	King Wizard	National
Kay-Bee	White	Kingsford	Goodrich
Kearl's Special	A. G. Mason Mfg. Co.	Kingston	National
Keasler	White	Kingsville	A. G. Mason Mfg. Co.
Keech	Free	Kion San Son	National
Keen-Edge	Davis	Kipp	A. G. Mason Mfg. Co.
Keen Klipper	Standard	Kirby	A. G. Mason Mfg. Co.
Keesling	White	Kirk	A. G. Mason Mfg. Co.
Keiser	A. G. Mason Mfg. Co.	Kirkcaldies	Standard
Kellers	A. G. Mason Mfg. Co.	Kirkman	A. G. Mason Mfg. Co.
Kelly	Davis	Kirkpatrick	A. G. Mason Mfg. Co.
Kelly, S. R.	White	Kirk, The	Standard
Kempisky	Standard	Kirkwood	Goodrich
Kendrick	Davis	Kitchener	White
Kendrick Special	A. G. Mason Mfg. Co.	Kitrell Gem	A. G. Mason Mfg. Co.
Kenmore	Standard	Klains Special	A. G. Mason Mfg. Co.
Kennedy	Standard	Klein	A. G. Mason Mfg. Co.
Kennedy	A. G. Mason Mfg. Co.	Klotes	A. G. Mason Mfg. Co.
Ken Quality	New Home	Knabe	Davis
Kensington	National	Knapps	Goodrich
Kensley	A. G. Mason Mfg. Co.	Kneuaman	A. G. Mason Mfg. Co.
Kent	Goodrich	Knickerbocker	Free
Kent	A. G. Mason Mfg. Co.	Kniffens	A. G. Mason Mfg. Co.
Kenton	Davis	Knight	A. G. Mason Mfg. Co.
Kentucky	National	Knor II	Standard
Kentucky	Standard	Knox	Free
Kentucky Beauty	Standard	Knox Special	White
Kentucky Gem	National	Kobacker	National
Kentucky Queen	National	Koenic	White
Kenwood	National	Kokoya	Standard
Kenwood	Standard	Kolgina	Free
Kenyon, M. S.	Standard	Konitzky	Free
Kerris Special	A. G. Mason Mfg. Co.	Kooken, J. T.	A. G. Mason Mfg. Co.
Kerr Kesterson	White	Korsmeyer, The	New Home
Keystone	A. G. Mason Mfg. Co.	Kracker Jack	A. G. Mason Mfg. Co.
Kibbers	Goodrich	Krafts Gem	National
Kiesee	White	Kragens	Goodrich
Kilgore	A. G. Mason Mfg. Co.	Kreas	A. G. Mason Mfg. Co.
Killian B	Standard	Krecic	White

Sewing-Machine Name	Sewing-Machine Company	Sewing-Machine Name	Sewing-Machine Company
Kreigers	National	La Favorite	Standard
Kroegers	National	La Foliscienso	Standard
Krogers Special	Free	La Francis	Standard
Krogers	National	La Francesca	Standard
Kuehler	A. G. Mason Mfg. Co.	La Francia Maritima	New Home
Kuehler Kuenman	White	La Frontora	National
Kvinden oz Hyemonet	New Home	La Garalda	National
Kvinden Pride	A. G. Mason Mfg. Co.	La Grand	National
Ky Bell	Free	La Granda	National
		La Homadien	National
La Argentine	A. G. Mason Mfg. Co.	La Hondurena	Goodrich
La Aurora	Standard	La Ideal	National
La Belle	Free	La India	National
La Belle	National	La International	National
La Mexicana	National	La Jalisciense	Standard
La Merocha	Standard	La Joya	A. G. Mason Mfg. Co.
La Moderna	Goodrich	La Joya-del-Hager	New Home
La Nacional	National	La Librias	New Home
La Novedades	National	La Ludo	National
La Nuevadades	National	La Manabela	National
La Opera	Standard	La Mariposa	New Home
La Palma	National	La Marquez	Standard
La Panamania	Davis	La Marquise	Free
La Pavoveedora	Davis	La Masivilla	National
La Perfection	Standard	La Mejor	Standard
La Boila Cinfuega	National	La Meriyana	Standard
La Bonita	Goodrich	La Meriposa	New Home
La Borda	A. G. Mason Mfg. Co.	La Perfection	Standard
La Bordadora	A. G. Mason Mfg. Co.	La Perfection	A. G. Mason Mfg. Co.
La Borla	A. G. Mason Mfg. Co.	La Perla	A. G. Mason Mfg. Co.
La Canadian	Davis	La Perlu-la-Casa	National
La Capoise	Standard	La Platta	A. G. Mason Mfg. Co.
La Casa Grandoa	A. G. Mason Mfg. Co.	La Princess	A. G. Mason Mfg. Co.
La Charmont	A. G. Mason Mfg. Co.	La Pueblo	National
La Chricanita	New Home	La Ramha	A. G. Mason Mfg. Co.
La Cinfiango	Standard	La Regent	Free
La Compagngo	Standard	La Reina	A. G. Mason Mfg. Co.
La Compedora	A. G. Mason Mfg. Co.	La Rosa	A. G. Mason Mfg. Co.
La Competida	National	La Rose	Free
La Constancia	Standard	La Saporita	National
La Costrera	Standard	La Silencosa	New Home
La Dalescieura	Standard	La Simpatica	Standard
La Ducheese	Free	La Sin Bomb	National
La Ecquador	National	La Sin Rinal	National
La Ecquadorator	National	La Soporeso Pueblo	National
La Esmeraldo	Goodrich	La Tarapaco	National
La Especatz	A. G. Mason Mfg. Co.	La Ultima	Standard
La Especial	Standard	La Valcedora	Standard
La Ertrella	National	La Velox	A. G. Mason Mfg. Co.
La Fabricas Universales	New Home	La Vercost	National
La Facilla	Goodrich	La Viebriosa	National
La Famillar	New Home	Lacey	A. G. Mason Mfg. Co.
La Favorite	National	Laclede	National

Sewing-Machine Name	Sewing-Machine Company	Sewing-Machine Name	Sewing-Machine Company
Ladies' Comfort	National	Leeth	A. G. Mason Mfg. Co.
Ladies Favorite	Standard	Leggett	Standard
Ladies Friend	National	Leigh	National
Ladies Home Journal	National	Leinburg	A. G. Mason Mfg. Co.
Lady Baltimore	Davis	Leiss	A. G. Mason Mfg. Co.
Lady Gay	National	Leland	Free
Lafayette	A. G. Mason Mfg. Co.	Lemberg	A. G. Mason Mfg. Co.
Laideau	Standard	Lenhart	Goodrich
Lake	A. G. Mason Mfg. Co.	Lennington	National
Lakeshore	National	Lenora	Davis
Lakeside	Free	Lenox, The	National
Lakeview	Davis	Leonard	A. G. Mason Mfg. Co.
Lakewood	Goodrich	Leopold Loeb	A. G. Mason Mfg. Co.
Lamar	Free	Lerrington	Standard
Lamberg	A. G. Mason Mfg. Co.	Leslie, The	National
Lamback	A. G. Mason Mfg. Co.	Lester	White
Lancaster	Standard	Leuerett	White
Lancaster	A. G. Mason Mfg. Co.	Levy Bros.	White
Landers	A. G. Mason Mfg. Co.	Levys	Standard
Landrys	A. G. Mason Mfg. Co.	Levis	Standard
Lanford	A. G. Mason Mfg. Co.	Lewis	White
Langs	Goodrich	Lexington	Davis
Lanham	A. G. Mason Mfg. Co.	Lexington	Standard
Lanston	A. G. Mason Mfg. Co.	Lexington	A. G. Mason Mfg. Co.
Lankford	A. G. Mason Mfg. Co.	Liberty	Davis
Lapiers	Goodrich	Liberty	National
Lark	Davis	Liberty Bell	Free
Lark	Free	Liberty Bell	National
Lashley	A. G. Mason Mfg. Co.	Liberty Rotary	White
Laubacks	A. G. Mason Mfg. Co.	Lichter Gem	National
Laurance	White	Lightfoot	A. G. Mason Mfg. Co.
Laurel	Free	Lighting	National
Laurence	Free	Lillard	A. G. Mason Mfg. Co.
Laval	Davis	Lima House	White
Laverty	A. G. Mason Mfg. Co.	Limestone	White
Lavina	Free	Limited	Free
Law	A. G. Mason Mfg. Co.	Lincoln	National
Lawfer	Davis	Lindake	National
Laws	White	Lindel	National
Lawrence	Free	Lindenheim	Standard
Lawthers	White	Lindheim	Standard
Lazarus Rotary	White	Lindholmes	National
Leader, New	New Leader	Linnator	Goodrich
Leader, New Century	New Leader	Lion	National
Leader, Our	National	Lipscomb	White
Leader, The	Standard	Lit's Rotary	White
Leath	White	Little Gem	National
Leauell	White	Little Wonder	New Home
Leavell	A. G. Mason Mfg. Co.	Little Worker	New Home
Lebecks	Standard	Litts	White
Lechles	Standard	Liquid Cool	Free
Lees	Davis	Livestock	Goodrich
Lees	Goodrich	Livingstone	National

Sewing-Machine Name	Sewing-Machine Company	Sewing-Machine Name	Sewing-Machine Company
Llewllous	White	Luth J. Y. Cia Sue	National
Lloyd	White	Lybrando	White
Lockman	Unknown	Lyle	A. G. Mason Mfg. Co.
Locke	Standard	Luzon	A. G. Mason Mfg. Co.
Loeb	A. G. Mason Mfg. Co.		
Loeser	Davis	Mab	Goodrich
Lowenstein	A. G. Mason Mfg. Co.	Mable	A. G. Mason Mfg. Co.
Loffers	A. G. Mason Mfg. Co.	Macey's Electric	National
Loflin	White	Macey's Leader	New Home
Loflin Bol.	White	Maceys New Empire	New Home
Logans Rotary	National	Maceys R. R. Red Star	Davis
Logna Rotary	National	Maceys Special	New Home
Lomax	A. G. Mason Mfg. Co.	Mackiss	White
Lonard	National	Macks	White
Lonestar	Davis	Macon	A. G. Mason Mfg. Co.
Lone Star A, B, C, D	Free	Macy	Davis
Long	National	Macy's Rotary	White
Longin	A. G. Mason Mfg. Co.	Madagascar	White
Longmire	White	Maddington	A. G. Mason Mfg. Co.
Longs	Standard	Maddox	A. G. Mason Mfg. Co.
Longs Favorite	Free	Madison	Standard
Longtino	White	Madison	National
Lonkes Special	National	Magill	White
Lopez Pewa Cia	Standard	Magnet	A. G. Mason Mfg. Co.
Lorain	A. G. Mason Mfg. Co.	Magnolia	National
Lorch	National	Magnolia	Standard
Lorena	National	Mail & Breeze	National
Loretta	Davis	Maine	Standard
Lotions	Standard	Majestic	New Home
Lottie	Standard	Major	Goodrich
Lott-W-Co.	White	Malba	A. G. Mason Mfg. Co.
Lott-Walker Co.	A. G. Mason Mfg. Co.	Malcohns	Standard
Louise	National	Malcolm	Standard
Louisiana	Free	Malones	Standard
Lowe	A. G. Mason Mfg. Co.	Maloness	White
Lowenberg	White	Malta	White
Lowenstein	A. G. Mason Mfg. Co.	Maltnomer	National
Lowett	A. G. Mason Mfg. Co.	Manavata	Standard
Lowey	Goodrich	Manawater	Standard
Lowman	White	Manchester	Goodrich
Lozier	National	Mandels	National
Lubra	Goodrich	Manfield	National
Lucas	National	Manhattan	Davis
Lucas	White	Manhattan	National
Luce	A. G. Mason Mfg. Co.	Manhattan	Standard
Lucerne	White	Manilla	National
Lucille	National	Manistee	White
Luckey Platt Co.	National	Manix	Standard
Lugers	White	Manlfair	White
Lugon	White	Manlove	A. G. Mason Mfg. Co.
Lula	A. G. Mason Mfg. Co.	Mansifield	Free
Lumpkin	A. G. Mason Mfg. Co.	Mansfield	White
Lunsford	White	Mansion	Davis

Sewing-Machine Name	Sewing-Machine Company	Sewing-Machine Name	Sewing-Machine Company
Mansion	National	Mayfield	A. G. Mason Mfg. Co.
Manter	Goodrich	May Flower	Goodrich
Manwestlake	New Home	Mayhen Lbr. Co.	White
Maple	White	May Queen	Goodrich
Maple Leaf	Standard	Mays Special	White
Marathon	Davis	Maywood	Goodrich
Maravana	Standard	McAlpin	Davis
March	White	Mc Alpin	White
March, J. B.	A. G. Mason Mfg. Co.	McAnish	National
Marchona	A. G. Mason Mfg. Co.	McCall	A. G. Mason Mfg. Co.
Marcus	Free	McCallum's Rotary	White
Marcy	Unknown	McCalm	A. G. Mason Mfg. Co.
Margareta	Standard	McCalvin	A. G. Mason Mfg. Co.
Margets	Standard	McCarthy	A. G. Mason Mfg. Co.
Marguerite	Household	McCaslin	A. G. Mason Mfg. Co.
Mariana Furniture Co.	A. G. Mason Mfg. Co.	McChesney	A. G. Mason Mfg. Co.
Marilton	Goodrich	McClung A, C, D	Goodrich
Marka Terbie	A. G. Mason Mfg. Co.	McCormick	Standard
Markham	White	McCracken	A. G. Mason Mfg. Co.
Mars	New Home	McCuiston	White
Marse	White	McCurdy	White
Marshall	A. G. Mason Mfg. Co.	McDaniel	A. G. Mason Mfg. Co.
Marshall Field	Standard	McDermot	Standard
Marshall Wells	National	McDonald	Standard
Martha	A. G. Mason Mfg. Co.	McDonald	A. G. Mason Mfg. Co.
Martha Washington	White	McDowell	National
Mart Heidrich	Goodrich	McEvans	White
Martin	Free	McGill	A. G. Mason Mfg. Co.
Martin	National	McGlynns	White
Marvel	National	McGraths	A. G. Mason Mfg. Co.
Marvel	Standard	McGreery	New Home
Mary, A. M. and Son	A. G. Mason Mfg. Co.	McGuiston	A. G. Mason Mfg. Co.
Maryland	National	McKay	A. G. Mason Mfg. Co.
Mascot	New Home	McKeey, The	Unknown
Mascota	Davis	McKelvey	Free
Mason	A. G. Mason Mfg. Co.	McKinley	A. G. Mason Mfg. Co.
Mason Rotary	A. G. Mason Mfg. Co.	McKinney	Unknown
Mason Rotary	White	McKissack	National
Mason Special	A. G. Mason Mfg. Co.	McLain's Best	A. G. Mason Mfg. Co.
Masonia	A. G. Mason Mfg. Co.	McLarty	White
Master	National	McLeods Store	A. G. Mason Mfg. Co.
Master, The	Standard	McMeans Bride	A. G. Mason Mfg. Co.
Matchless	Davis	McMeans Pride	A. G. Mason Mfg. Co.
Matchless	A. G. Mason Mfg. Co.	McMitch	National
Matchless	Standard	McMillen	A. G. Mason Mfg. Co.
Mathews	Davis	McNeil	A. G. Mason Mfg. Co.
Matron, The	National	McNicols	National
Maud	Davis	McPetters	White
Maxfield	White	McRae	A. G. Mason Mfg. Co.
Maxwell	White	McReas	Davis
May and Graix	White	McSwean	A. G. Mason Mfg. Co.
May Company, The	Standard	McVay Special	A. G. Mason Mfg. Co.
Mayfair	White	McWherr	White

Sewing-Machine Name	Sewing-Machine Company	Sewing-Machine Name	Sewing-Machine Company
McWilliams	A. G. Mason Mfg. Co.	Minges	Standard
Meckes	Free	Mine Door	A. G. Mason Mfg. Co.
Medon	Unknown	Mine Dor	White
Meisers	White	Miner	A. G. Mason Mfg. Co.
Meldrum	National	Minerva	Standard
Meller	National	Minneapolis	National
Meller	Unknown	Minneapolis	Standard
Melone	Davis	Minnehaha	National
Melone	Unknown	Minnehaha	Standard
Melrose	National	Minnesota	Davis
Melville	National	Minnesota H	Standard
Melvin	White	Minnesota 1	New Home
Menda	Standard	Minnesota A9-3500000	Davis
Menelee	A. G. Mason Mfg. Co.	Minnesota 2 & 3	Davis
Menough	Standard	Minnie, New	A. G. Mason Mfg. Co.
Merchant & E	A. G. Mason Mfg. Co.	Miracle	Davis
Mercantile	Standard	Mirror and Farmer	Standard
Mercury	Goodrich	Mission	Davis
Meridan	Standard	Missouri	National
Merrell	Free	Mississippi	White
Merrick	Standard	Mistree	Standard
Mervin, K., Co.	White	Mistress	Standard
Mesco	New Home	Mitchell	Standard
Messayen	A. G. Mason Mfg. Co.	Mitchell	A. G. Mason Mfg. Co.
Metcalf	A. G. Mason Mfg. Co.	Mi-Well Co.	White
Meteor	New Home	Moberly	Standard
Metropolis	Standard	Mobery	Standard
Metropolitan	National	Mobile	Davis
Mexbo	Free	Model	Davis
Meyer	A. G. Mason Mfg. Co.	Model	Goodrich
Mhoon	White	Model D	Davis
Miami Mer. Co.	A. G. Mason Mfg. Co.	Model Queen	A. G. Mason Mfg. Co.
Michel	Free	Model Sewer	Standard
Michigan	New Home	Modern	Davis
Michigan	White	Modern	A. G. Mason Mfg. Co.
Michigan Farmer	Davis	Modern Wilson	A. G. Mason Mfg. Co.
Midkiff	White	Modern Woodsman	Goodrich
Midland	Davis	Modes	National
Midland	National	Modiste	Free
Midway	White	Moerbe	White
Millard	Standard	Moerle	A. G. Mason Mfg. Co.
Mikady	Standard	Mohand	Davis
Milam	White	Mohawk	National
Millane	Standard	Moiser	White
Millard	Standard	Moister	A. G. Mason Mfg. Co.
Miller	Standard	Mollen Kamp	A. G. Mason Mfg. Co.
Miller	White	Moll Special	White
Miller & W	National	Moms	National
Milligan Cash	A. G. Mason Mfg. Co.	Momas	National
Milton	Free	Monarch	Davis
Milwauka	Free	Monarch	New Home
Milwauka	Goodrich	Modys Favorite	Standard
Minakif Leader	A. G. Mason Mfg. Co.	Money King	Standard

Sewing-Machine Name	Sewing-Machine Company	Sewing-Machine Name	Sewing-Machine Company
Monitor	Davis	Nacional Standard	Standard
Monitor	National	Nadbers	White
Monogram	Davis	Nagorski	White
Monroe	Davis	Namson	Davis
Monte-go	White	Namson	A. G. Mason Mfg. Co.
Montejo Rotary	White	Nance	A. G. Mason Mfg. Co.
Montemayor	New Home	Napier	Standard
Montgomery	Standard	Nash	White
Montgomery Ward	National	Nashville	Goodrich
Moores	A. G. Mason Mfg. Co.	Nath Mf. and Hub. Co.	White
Morehouse	Davis	National A,B,C,D,E,F,G,H,O	National
Morey	Standard	National Baptistiec	National
Morgan	Goodrich	National Home	Free
Morley	A. G. Mason Mfg. Co.	National Rotary	National
Morris	A. G. Mason Mfg. Co.	National Supply Co.	National
Morrison	Davis	Nation's Pride	Standard
Morrison	A. G. Mason Mfg. Co.	Naura India	National
Morrison Bros.	Standard	Navarre	Davis
Morse	Free	Naylors Special	A. G. Mason Mfg. Co.
Morton	White	Nebraska	Free
Morum Bros.	New Home	Neely and C	A. G. Mason Mfg. Co.
Mosby	A. C. Mason Mfg. Co.	Nellie	National
Moser	White	Nelson	New Home
Moss	A. G. Mason Mfg. Co.	Nesbit	National
Mountaineer	Standard	Netzon	Davis
Moyer Special	A. G. Mason Mfg. Co.	Neurva National	New Home
Moyers Pride	Standard	Neva	Standard
Moyles Special	A. G. Mason Mfg. Co.	Neversuch	Standard
Mt. Calm	A. G. Mason Mfg. Co.	Neville	A. G. Mason Mfg. Co.
Muelder	White	New Acme	Davis
Mulcock, J. K.	A. G. Mason Mfg. Co.	New Acme	Standard
Mulder Jewell	A. G. Mason Mfg. Co.	New Ane	National
Mulleins	A. G. Mason Mfg. Co.	New Albany	National
Mullins	Davis	New Ally	National
Mund	Domestic	New American	Davis
Mundo de Colon	National	New American Lichlador	National
Munn	A. G. Mason Mfg. Co.	New Atlas	Standard
Munroe	A. G. Mason Mfg. Co.	New Automatic	Free
Murphys	Standard	New Avonia	New Home
Murphys and Allen	A. G. Mason Mfg. Co.	New Bach	White
Murray	A. G. Mason Mfg. Co.	New Belle	National
Murray, The	Standard	New Belle	Standard
Muskingdom	Standard	New Ben Hur	National
Muskoka	Goodrich	New Bevon	Free
Mussina	National	New Bouquett	National
Myers	A. G. Mason Mfg. Co.	New Cambridge	New Home
Myers	Standard	New Carson	New Home
My Own	A. G. Mason Mfg. Co.	New Centenial	Davis
My Own	Standard	New Century	Standard
Myrtle	Goodrich	New Century	A. G. Mason Mfg. Co.
		New Century Leader	New Leader
Nabob	Davis	New Century Rotary	White
Nacional	National	New Champion	New Home

Sewing-Machine Name	Sewing-Machine Company	Sewing-Machine Name	Sewing-Machine Company
New Chronicle	National	New Home 400	New Home
New Colonial House	Davis	New Home Queen	New Home
New Columbia	New Home	New Home Rotary	New Home
New Columbus	Standard	New Home Star	New Home
New Comer	White	New Homestead	Free
New Companion	New Home	New Homestead	Standard
New Concord	Standard	New Home Superior	New Home
New Conover	Davis	New Home Victoria	New Home
New Cottage	Davis	New Hope	Davis
New Cottage	National	New Household	National
New Damascus	A. G. Mason Mfg. Co.	New Howard	Free
New Daisey	A. G. Mason Mfg. Co.	New Howe	Standard
New Davis	Davis	New Idea	A. G. Mason Mfg. Co.
New Decatur	National	New Ideal	New Home
New Defender	White	New Idemascus	White
New Departure	National	New Jewell	National
New Dixie	Davis	New King	National
New Domestic	Domestic	New Leader	New Leader
New Devon	National	New Lennox	Standard
New Duplex	Free	New Life	National
New Easy Running	National	New Love	New Leader
New Elgin	A. G. Mason Mfg. Co.	New Mason	White
New Elmore	National	New Mason's	White
New Empire	National	New Matchless	Standard
New Empress	National	New Matthews	A. G. Mason Mfg. Co.
New England	Davis	New May	A. G. Mason Mfg. Co.
New England	A. G. Mason Mfg. Co.	New Merrick	A. G. Mason Mfg. Co.
New England Queen	National	New Method	Standard
New England Rotary	National	New Method	A. G. Mason Mfg. Co.
New Equity	National	New Mexicano	Standard
New Era	National	New Mode	National
New Ermine	A. G. Mason Mfg. Co.	New Model	Goodrich
New Family	New Home	New Monitor	Free
New Favorite	Davis	New Monroe	A. G. Mason Mfg. Co.
New Fillespie	National	New Moulton	Free
New Fireside	Davis	New More	Standard
New Fireside	National	New National	New Home
New Firm	Standard	New National	White
New Florence	A. G. Mason Mfg. Co.	New Netzow	Davis
New Florence Rotary	A. G. Mason Mfg. Co.	New Oriole	National
New Gold Dust	Davis	New Oswald	A. G. Mason Mfg. Co.
New Goodrich	Goodrich	New Pacific	National
New Gourley	Davis	New Paguet	National
New Havana	Standard	New Palace	A. G. Mason Mfg. Co.
New Havan	New Home	New Parish	Standard
New Haven	New Home	New Peerless	Standard
New Haven Imp.	Standard	New Peoria	National
New Hibbare	Davis	New Perfection	A. G. Mason Mfg. Co.
New Hickory	Davis	New Pillsbury	National
New Holmes	Standard	New Pittsburgh B	Goodrich
New Holmes	White	New Post	National
New Home	New Home	New Premium	National
New Home Central	New Home	New Presto	National

Sewing-Machine Name	Sewing-Machine Company	Sewing-Machine Name	Sewing-Machine Company
New Principle	Free	New Wilson	A. G. Mason Mfg. Co.
New Progress	Davis	New Wilsons	A. G. Mason Mfg. Co.
New Queen	A. G. Mason Mfg. Co.	New Winchester	New Home
New Rapid	Standard	New Winner	Davis
New Reliable	Davis	New Yale	Foley & Williams Mfg. Co.
New Rex	Davis		
New Rival	Goodrich	New York	Goodrich
New Rome	National	New York	White
New Rotary	White	New York Gem	A. G. Mason Mfg. Co.
New Royal	Free	New Z Pasquet	National
New Royal A	Free	Newark	Goodrich
New Royal Queen	Free	Newman	White
New Single	A. G. Mason Mfg. Co.	Newport	Standard
New Sinning	A. G. Mason Mfg. Co.	Newton	Davis
New Sincer	New Home	Newton	Free
New South 1	Davis	Newton	Standard
New South 2	Goodrich	Niagra	National
New South 3	National	Nichols	A. G. Mason Mfg. Co.
New Star	Davis	Nicholson	Standard
New Star	Standard	Nicholson	White
New Star	A. G. Mason Mfg. Co.	Niobe	New Home
New States	Free	Nicce	White
New Sterling	Free	Nimmons	White
New Steward	Standard	Nixon	A. G. Mason Mfg. Co.
New Stockman	Davis	Nofzinger	A. G. Mason Mfg. Co.
New Success	Standard	No Name	National
New Superb	Davis	Nonequat	White
New Treasure	A. G. Mason Mfg. Co.	Nonpariel	National
New Tremont	National	Noonan	A. G. Mason Mfg. Co.
New Trip	National	Nordica	Davis
New Triumph	Davis	Nordon	National
New Tuxedo	Standard	Norella	White
New Tyler	A. G. Mason Mfg. Co.	Norfolk	Standard
New Vance	Davis	Norfolk	White
New Vassar	Davis	Norleigh Diamond	Davis
New Victor	National	Norman	A. G. Mason Mfg. Co.
New Victoria	New Home	Norris	White
New Victory	Davis	North American	National
New Victory 9	Goodrich	Northcut	A. G. Mason Mfg. Co.
New Waltham	Goodrich	Northern	New Home
New Warwick	A. G. Mason Mfg. Co.	North Queen	National
New Way	Free	North Western	National
New Way Rotary, The	White	Norwood	Standard
New Weed	Standard	Not Nac	Davis
New White	White	No-Try-On	National
New White Peerless	A. G. Mason Mfg. Co.	Nova	National
New White Rotary	White	Nova Machine Roses	Standard
New Whiteley	National	Noxabee Queen	National
New Whitesley	A. G. Mason Mfg. Co.	Noxall	National
New Willard	White	Nox, X. L.	National
New Williams	Unknown	Nuerva	New Home
New Wilmot	Unknown	Nunnely	White
New Wilson	Davis	Nutflake	White

Sewing-Machine Name	Sewing-Machine Company	Sewing-Machine Name	Sewing-Machine Company
Ny Luchador	National	Orchid	Davis
		Orchid Electric	A. G. Mason Mfg. Co.
Oakdale	Standard	Ordway	New Home
Oakland	Goodrich	Oregon	Goodrich
Oakwood	Free	Orient	National
Oakwood, A. B.	Davis	Orient	Standard
Oberlin	Davis	Oreint	White
Observer	Goodrich	Oreole	Standard
Occident	National	Oretan	National
Occental	National	Oritana	National
Odelbury	White	Orleans	Standard
Oden	A. G. Mason Mfg. Co.	Orme Co.	A. G. Mason Mfg. Co.
Oettinger	White	Osborn	Unknown
Ogle	National	Osborn	A. G. Mason Mfg. Co.
Ogle, B.	White	Oscar Hewett Rotary	White
Oglethorpe	Free	Ostertag	A. G. Mason Mfg. Co.
Ohio	Davis	Osvetta	A. G. Mason Mfg. Co.
Ohio Farmer	Foley & Williams	Oswale	A. G. Mason Mfg. Co.
Ohio Farmer	Goodrich	Ottowa	Household
Ohio Farmer	Standard	Otto	Standard
Ohio Farmer Model D	Davis	Otto Klein	A. G. Mason Mfg. Co.
Ohio Valley	Davis	Ouatonna	White
Okeh	Davis	Our Bonanza	A. G. Mason Mfg. Co.
Okeh	New Home	Our Gem	White
Oklahoma	Davis	Our Grand	Davis
Oklahoma, The	Free	Our Leader	Free
Old Homestead	National	Our Model	Standard
Old Homestead Rotary	National	Our Own	Free
Old Reliable	A. G. Mason Mfg. Co.	Our Pet	National
Oldenberg	A. G. Mason Mfg. Co.	Our Pride	National
Olds Worthman Co.	Standard	Our Salesman	Free
Oldson Leader	A. G. Mason Mfg. Co.	Our Special	Free
Olive, The	A. G. Mason Mfg. Co.	Our Special	Goodrich
Olney	Goodrich	Our Surprise	National
Olney	Standard	Our Troveler	National
Olympia	National	Our Very Best	Standard
Olympia	Standard	Our Wonder	Goodrich
Omaha	Free	Ousguard and V.	White
Omega Rotary	White	Outlet Special	National
Onachita Bell	Davis	Overstreet	White
Oneida	National	Owattona	A. G. Mason Mfg. Co.
O'Neil Pride	A. G. Mason Mfg. Co.	Oxford	White
O'Neil, The	Standard	Oxford 1	Goodrich
O'Neil Rotary	White	Oxford 2	National
Ongnard	A. G. Mason Mfg. Co.	Oxford Singer	Unknown
Onondazo	Free	Ozark	Free
Ontarri	National		
Ontario	A. G. Mason Mfg. Co.	Pace	A. G. Mason Mfg. Co.
Oppemheimer	National	Pacific	National
Oracle	New Home	Pacific	Standard
Orade	New Home	Pacific Queen	Goodrich
Orange	Davis	Paddock	Davis
Oransky	A. G. Mason Mfg. Co.	Page	A. G. Mason Mfg. Co.

Sewing-Machine Name	Sewing-Machine Company	Sewing-Machine Name	Sewing-Machine Company
Pagoma	Davis	Pearl	Standard
Palace	National	Pearlhaven	Standard
Palcios and Zambreno	A. G. Mason Mfg. Co.	Pearline	National
Palma	A. G. Mason Mfg. Co.	Pecko Special	A. G. Mason Mfg. Co.
Palmer	National	Peco Electric	New Home
Palmetto	A. G. Mason Mfg. Co.	Pee Dee	A. G. Mason Mfg. Co.
Pan American	New Home	Pee-Gee	Davis
Pansor	National	Peepless	White
Panton and Whit	Goodrich	Peepp	White
Paragold	A. G. Mason Mfg. Co.	Peeress	Standard
Paragon	Davis	Peerless	Free
Paragon	Standard	Peerless	National
Paragon, The New	Standard	Peerless Rotary	National
Paramount	Davis	Peerless Rotary	A. G. Mason Mfg. Co.
Paramount	Standard	Pelantense	White
Parchman	A. G. Mason Mfg. Co.	Pelham	National
Partridge	A. G. Mason Mfg. Co.	Pelican	Free
Parent	A. G. Mason Mfg. Co.	Pelican	Standard
Parinas	Davis	Pelleter	Free
Paris	A. G. Mason Mfg. Co.	Pellerin	National
Paris King	A. G. Mason Mfg. Co.	Pelotence	A. G. Mason Mfg. Co.
Parish Grand	Standard	Pelton	A. G. Mason Mfg. Co.
Parisian	Davis	Pemberton	Davis
Park City	National	Peninsular	National
Parker	National	Penitang	National
Parker	White	Penman	Standard
Parks	Standard	Penmam, H. D.	A. G. Mason Mfg. Co.
Parlor City	Davis	Penn	A. G. Mason Mfg. Co.
Parlor City	White	Penn, The	A. G. Mason Mfg. Co.
Parmao	White	Pennsylvania	Goodrich
Passaic	White	Penny	Davis
Passmore	A. G. Mason Mfg. Co.	Pensylvania	Goodrich
Passmore Rotary	White	People, The	Standard
Pastime	Standard	Peoples Furniture Company	A. G. Mason Mfg. Co.
Pate	White	Peoria	Davis
Pates	New Home	Peoria	New Home
Pates Special	Standard	Perer	White
Pathfinder	Davis	Perfect	A. G. Mason Mfg. Co
Pathway	Free	Perfecta, The	Standard
Pat-Mac	Davis	Perfection	Davis
Patrick's Gem	A. G. Mason Mfg. Co.	Perfection	Free
Pattelos Special	A. G. Mason Mfg. Co.	Perfection	Standard
Pattilo	White	Perfection	White
Patten	Goodrich	Perfecto	Goodrich
Paul	Davis	Perkins	White
Paul	National	Perla-de-la-Casa	Standard
Paul	A. G. Mason Mfg. Co.	Perpetual	National
Paveway	National	Perrless Rotary	White
Pea-Body	Davis	Perry and Henry	A. G. Mason Mfg. Co.
Peabody	A. G. Mason Mfg. Co.	Pet	Davis
Peach Belt	A. G. Mason Mfg. Co.	Pet	Free
Pearl	Davis	Pettis	Davis
Pearl	New Home	Pettis	Free

Sewing-Machine Name	Sewing-Machine Company	Sewing-Machine Name	Sewing-Machine Company
Petty Brothers	White	Popular	Davis
P. E. Victor	Free	Porch Privetts	White
Pfeuffer, H., Co.	White	Portage	A. G. Mason Mfg. Co.
Pfleger	A. G. Mason Mfg. Co.	Porter	Davis
Phar Bros.	A. G. Mason Mfg. Co.	Post Electric	National
Phelans	White	Potts, W. A.	A. G. Mason Mfg. Co.
Philadelphia	A. G. Mason Mfg. Co.	Potzer	Davis
Phila-Singer	Unknown	Powell	A. G. Mason Mfg. Co.
Phillip	A. G. Mason Mfg. Co.	Practical Farmer	Standard
Phillips	Standard	Praire Queen	Free
Phillippine	Davis	Prairie Queen	Free
Phillopene	Davis	Pratts	A. G. Mason Mfg. Co.
Phoenix	National	Premier	Davis
Phoenix	White	Premier	New Home
Pickering	Standard	Premier	National
Piedmont	Davis	Premier	Standard
Piedmont	Goodrich	Premium Chronicle	Goodrich
Piedmont Electric	New Home	Premium Chronicle, New	National
Pierce	White	Premium Dallas	Unknown
Pierce Gold	Standard	Premium Galveston	National
Piffer Special	A. G. Mason Mfg. Co.	Premium Houston	Goodrich
Pilgrims	National	Prescot Hardware Co.	A. G. Mason Mfg. Co.
Pillsbury	National	Presidentt	Davis
Pilot	Standard	Press Pub. Assn.	Standard
Pilsbury	National	Presto	A. G. Mason Mfg. Co.
Pinnens Pride	Free	Prewitt	A. G. Mason Mfg. Co.
Pinnic Kinick	Davis	Price and Allen	A. G. Mason Mfg. Co.
Pinnock	Goodrich	Priscilla	New Home
Pinnock	Standard	Pride of Home	A. G. Mason Mfg. Co.
Pioneer	National	Pride of Kentucky	Standard
Pioneer	Standard	Pride of Milwaukee	Free
Pioneer Press	National	Pride of Poteau	A. G. Mason Mfg. Co.
Piper	Standard	Pride of Virginia	A. G. Mason Mfg. Co.
Piquett	Goodrich	Pride of World	A. G. Mason Mfg. Co.
Pittman	A. G. Mason Mfg. Co.	Primmer	A. G. Mason Mfg. Co.
Pitts	White	Primor	New Home
Pittsburgh Electric	A. G. Mason Mfg. Co.	Prince	A. G. Mason Mfg. Co.
Plain	National	Princess	National
Plainview	White	Princess	Standard
Planet	Goodrich	Princeton	National
Planters	Standard	Principle	Goodrich
Planters	A. G. Mason Mfg. Co.	Pringle	Davis
Platt	Davis	Printzlaff	Free
Plymouth	Davis	Priscella	Davis
Plymouth	National	Priscilla	Unknown
PlyQueen	White	Priscilla	New Home
Polk and Co.	A. G. Mason Mfg. Co.	Pritzlaff	Davis
Polly	Standard	Prize Model	Goodrich
Polytype	National	Prize, The	National
Pointer Gem	A. G. Mason Mfg. Co.	Process	National
Pommer	National	Progress	Standard
Popes Queen	A. G. Mason Mfg. Co.	Progress	A. G. Mason Mfg. Co.
Poplar	National	Prospect	A. G. Mason Mfg. Co.

Sewing-Machine Name	Sewing-Machine Company	Sewing-Machine Name	Sewing-Machine Company
Prosperity	White	Queen White Sewing Machine Co.	A. G. Mason Mfg. Co.
Protens	National	Queen of the West	National
Providence	National	Queen Winner	A. G. Mason Mfg. Co.
Provodero del Hagar	A. G. Mason Mfg. Co.	Queen 12	Free
Public Service	White	Quickel's Best	Standard
Pullman	A. G. Mason Mfg. Co.	Quickstitch	Standard
Punxsutawney	White	Quiller, P. R.	A. G. Mason Mfg. Co.
Purdin	Goodrich	Quinan Och Hemmet	New Home
Puregold	Free	Quincy	Davis
Puritan	National	Quincy	Goodrich
Puritan	White	Quinn	A. G. Mason Mfg. Co.
Purity	National	Quison	Standard
Pursero	White		
Putnam	A. G. Mason Mfg. Co.	Raders	A. G. Mason Mfg. Co.
Pyramid	Goodrich	Radiant	Standard
		Radio	Davis
Quackenbush	Davis	Radior	A. G. Mason Mfg. Co.
Quaker	National	Radior Rotary	White
Quaker City	Free	Radner Special	A. G. Mason Mfg. Co.
Quality	Davis	Raesford Hardware Co.	A. G. Mason Mfg. Co.
Quality	National	Railway	National
Quality	Standard	Rainbow	Free
Quapa Special	National	Rainer	White
Queen	Davis	Rainha	White
Queen	A. G. Mason Mfg. Co.	Rainwater	Goodrich
Queen American	National	Raja	National
Queen Anne	Standard	Raleigh	National
Queen Bee	National	Ralston B	A. G. Mason Mfg. Co.
Queen Bess	Davis	Ramand	National
Queen Bess	Free	Ramco Rotary	White
Queen City	Davis	Ramons	National
Queen City	National	Ramsdells Best	A. G. Mason Mfg. Co.
Queen City	Unknown		National
Queen Cottage	National	Ranberry	Goodrich
Queen Family	A. G. Mason Mfg. Co.	Randchier	A. G. Mason Mfg. Co.
Queen Fairest	National	Randville	A. G. Mason Mfg. Co.
Queen Home	Davis	Raney	A. G. Mason Mfg. Co.
Queen Kentucky	National	Ranger	Davis
Queen May	National	Rankin	A. G. Mason Mfg. Co.
Queen Modern	A. G. Mason Mfg. Co.	Ranks	A. G. Mason Mfg. Co.
Queen, New	A. G. Mason Mfg. Co.	R. D. Masayeh R.	White
Queen New England	A. G. Mason Mfg. Co.	Read, The	Free
Queen North	National	Read, The	Standard
Queen Tuxubee	National	Read Rotary	White
Queen Pacific	Standard	Read & Snyder	National
Queen Plymouth	A. G. Mason Mfg. Co.	Reading Special	A. G. Mason Mfg. Co.
Queen Royal	A. G. Mason Mfg. Co.	Reagan	White
Queen Southern	A. G. Mason Mfg. Co.	Reams	White
Queen Smith	A. G. Mason Mfg. Co.	Record	National
Queen Texas	A. G. Mason Mfg. Co.	Rectors	White
Queen Toranto	National	Red Buza	Free
Queen Western	National	Red Cross	Davis

Sewing-Machine Name	Sewing-Machine Company	Sewing-Machine Name	Sewing-Machine Company
Red Cross	Standard	Richland	Standard
Redford	Goodrich	Richland	A. G. Mason Mfg. Co.
Redgeton	National	Richmond	Davis
Red Star	National	Richmond A, B, C, D, E	Free
Read & Son	A. G. Mason Mfg. Co.	Richter	Free
Reed Special	Standard	Richters	Standard
Regal	Davis	Rickoff	White
Regal	A. G. Mason Mfg. Co.	Ridder	National
Regal B, C, D Rotary	National	Ridgeway	National
Regent	Free	Riegle, The	A. G. Mason Mfg. Co.
Regina	Goodrich	Rierson	A. G. Mason Mfg. Co.
Rehan	Davis	Rierson, W. and S.	White
Reich	A. G. Mason Mfg. Co.	Rightway	Standard
Reimer	White	Rikes	Davis
Reimer Special	A. G. Mason Mfg. Co.	Riley	A. G. Mason Mfg. Co.
Reina	Standard	Ripley Belle	A. G. Mason Mfg. Co.
Reliable	National	Ritchable	National
Reliable	Standard	Ritchdale	National
Reliable A	Free	Ritchie Hardware Co.	Standard
Reliance	Davis	Ritman	White
Reliance	Free	Rittenhouse	Free
Reliance	Standard	Ritter	Standard
Reliance	A. G. Mason Mfg. Co.	Ritters	Goodrich
Reliance Hardware Co.	New Home	Rival	Davis
Remaly	Standard	Rival	Free
Remington	National	Rival	Standard
Renan	A. G. Mason Mfg. Co.	Rival Sin	A. G. Mason Mfg. Co.
Renfreu	Davis	Riverside	National
Reno	Goodrich	Riverside Extra	Standard
Renown	National	Riverton	Household
Republic	Davis	Rivore	A. G. Mason Mfg. Co.
Republic	National	Roanoke	Davis
Reputation	National	Rober	Standard
Revalation	National	Roberta	National
Rev-O-Nue	National	Roberta	A. G. Mason Mfg. Co.
Revork	Davis	Robert Lee	White
Rex	National	Roberts	Goodrich
Rexford	National	Roberts	Standard
Reynolds	Free	Robertson	Goodrich
Reynolds	White	Robins	National
Reznoir Rotary	A. G. Mason Mfg. Co.	Robinson	A. G. Mason Mfg. Co.
Reznor Rotary	White	Rockford	Free
Rhenania	Goodrich	Rockford	Standard
Rhoda and Happe	Standard	Rock City	National
Rhodesia	Standard	Rockford	Free
Rhone Hardware Co.	A. G. Mason Mfg. Co.	Rockland	A. G. Mason Mfg. Co.
Rhyne Brothers	A. G. Mason Mfg. Co.	Roding	National
Rialto	National	Roeser	Standard
Rice, The	Free	Rogers	A. G. Mason Mfg. Co.
Rice's	White	Rogers, E. M.	Standard
Richard, J. D.	A. G. Mason Mfg. Co.	Roiz	Standard
Richard's	White	Roland, J. A.	A. G. Mason Mfg. Co.
Richardson	Standard	Roman	White

Sewing-Machine Name	Sewing-Machine Company	Sewing-Machine Name	Sewing-Machine Company
Rome	A. G. Mason Mfg. Co.	Rugby	Free
Rominger	A. G. Mason Mfg. Co.	Rule	A. G. Mason Mfg. Co.
Romish Special	A. G. Mason Mfg. Co.	Run Easy	National
Ronaldson	Standard	Rural	A. G. Mason Mfg. Co.
Ronans	White	Rush	A. G. Mason Mfg. Co.
Rood	Standard	Rusmark	Free
Roops	A. G. Mason Mfg. Co.	Russell	A. G. Mason Mfg. Co.
Roos	White	Russer	A. G. Mason Mfg. Co.
Roots	National	Ruther	White
Rose	National	Rutherford	White
Roselyn	A. G. Mason Mfg. Co.	Rygresco Rotary	A. G. Mason Mfg. Co.
Rosemary	National	Rygroco Rotary	White
Rosenbaum	Davis	Rylando	A. G. Mason Mfg. Co.
Rosenberg	A. G. Mason Mfg. Co.	Ryno, C. V.	A. G. Mason Mfg. Co.
Rosenburg and H	Standard		
Rosenblum	Free	Sabin	Davis
Rosendale	Standard	Sacket Studio	A. G. Mason Mfg. Co.
Rosenbush	A. G. Mason Mfg. Co.	Sackett	White
Rosenthal Gem	A. G. Mason Mfg. Co.	Sacremento Star	Goodrich
Rosenwald	A. G. Mason Mfg. Co.	Sadie	Standard
Rosenwater	Standard	Sadler Special	A. G. Mason Mfg. Co.
Ross	A. G. Mason Mfg. Co.	Saidon	Free
Rossman	White	St. Clair	Goodrich
Ross Perfection	Standard	St. Hyancinthe	Goodrich
Rossenback	Standard	St. Jacobs	Standard
Rotary Special	A. G. Mason Mfg. Co.	St. James	Standard
Rotative A	National	St. John Royal	Free
Roth Special	Standard	St. Lawrence	Standard
Rothchild	Standard	St. Louis	National
Rotiscillo	Free	Saint Marys	National
Rotoro	National	St. Paul	Goodrich
Rottenberg	Goodrich	Salt City	A. G. Mason Mfg. Co.
Rover	Standard	Salt Lake	Davis
Rowe Brothers	A. G. Mason Mfg. Co.	Saltman	National
Rowel	A. G. Mason Mfg. Co.	Salva Hos	National
Rowell	White	Samaornials	National
Rowland	A. G. Mason Mfg. Co.	Sambrooks	A. G. Mason Mfg. Co.
Rowland King Co.	Standard	Sampson	Standard
Rowley	White	Sampson	White
Royal	Free	Samuel	New Home
Royal Electric	Free	Samuels	A. G. Mason Mfg. Co.
Royal Leader	A. G. Mason Mfg. Co.	Sanchez M. Y. Hno	National
Royal New	Free	Sanders Co.	A. G. Mason Mfg. Co.
Royal Oak	Free	Sanderson Co.	A. G. Mason Mfg. Co.
Royal Palm	A. G. Mason Mfg. Co.	Sangaman	Davis
Royal Ruby, The	New Home	Sanger	National
Royal St. John, New	Free	Sanger Grand B	National
Royal Saint John O. S.	Free	Sans Favorite	Standard
Royal St. John O. S. H.	Free	Sarver and Spencer	A. G. Mason Mfg. Co.
Ruby	New Home	Saskatchewan	A. G. Mason Mfg. Co.
Rudolph A, B, C	Goodrich	Sancer Special	Free
Rudy	Free	Savage	National
Rugby	Davis	Savoy	A. G. Mason Mfg. Co.

Sewing-Machine Name	Sewing-Machine Company	Sewing-Machine Name	Sewing-Machine Company
Saxon Special	A. G. Mason Mfg. Co.	Selecta	Standard
Scandanavian	Goodrich	Selecto	White
Scar	A. G. Mason Mfg. Co.	Selma C., D., E.	Free
Scarborough	National	Selsers Pride	Standard
Schackeford	Standard	Semense	Standard
Schafer	A. G. Mason Mfg. Co.	Seminole	Free
Scheats	A. G. Mason Mfg. Co.	Seneca	Free
Scheets	A. G. Mason Mfg. Co.	Seneca Chief	A. G. Mason Mfg. Co.
Schelemer	A. G. Mason Mfg. Co.	Senorita	Standard
Schelnk	Standard	Sentinel	National
Schelton's Pride	A. G. Mason Mfg. Co.	Sequoia	Davis
Schertz	A. G. Mason Mfg. Co.	Serata	New Home
Scheusseers Queen	A. G. Mason Mfg. Co.	Serimshans	White
Schlemier	A. G. Mason Mfg. Co.	Seroco	National
Schmidt	A. G. Mason Mfg. Co.	Service	Davis
Schilz	National	Service	Free
Schmit & S.	A. G. Mason Mfg. Co.	Sewards	A. G. Mason Mfg. Co.
School-Rotschild	Davis	Sew Easy	A. G. Mason Mfg. Co.
Schoonmaker	National	Sew Easy	White
Schremie Co.	A. G. Mason Mfg. Co.	Sew Right	A. G. Mason Mfg. Co.
Schultz	White	Sewell	A. G. Mason Mfg. Co.
Schultz & Howe	A. G. Mason Mfg. Co.	Seybold	Davis
Schultze's Gem	A. G. Mason Mfg. Co.	Seybold Bros.	Standard
Schuman	A. G. Mason Mfg. Co.	Shamrock	Standard
Schwartz Kopp	National	Shapleigh	Davis
Scientific	Standard	Shapleigh	Standard
Scio	A. G. Mason Mfg. Co.	Sharon	Sharon, Sharon, Ohio (1907)
Scanavely, C. C.	National		
Scofield	A. G. Mason Mfg. Co.	Sharwick	Standard
Scott	Standard	Shawnee Chief	Davis
Scott L. G.	A. G. Mason Mfg. Co.	Sheffield	White
Scoville New Era	National	Shelby	Goodrich
Scruggs	Free	Sheldon	National
Scully	National	Sheller	White
Seagal Special	A. G. Mason Mfg. Co.	Shenley, The	National
Seal's Pride	Standard	Sheppard	Standard
Seamstress	Davis	Sheppard Special	A. G. Mason Mfg. Co.
Seamstress, New	New Home	Sheppars Favorite	Standard
Seamstress, Old	Goodrich	Sherman	Standard
Sea Side	A. G. Mason Mfg. Co.	Shettucket	Davis
Seasons 4	A. G. Mason Mfg. Co.	Shifley	A. G. Mason Mfg. Co.
Sears, H. D.	A. G. Mason Mfg. Co.	Shinks	White
Sears, Roebuck	National	Shipkonsky	Standard
Seaver's Special	A. G. Mason Mfg. Co.	Shipley and Boht	A. G. Mason Mfg. Co.
Seavoy	Standard	Shoneman	Standard
Secora	National	Shook and Shook	Davis
Secrocer	A. G. Mason Mfg. Co.	Shook and Shook	White
Security	Davis	Shordemanden	Goodrich
Security	A. G. Mason Mfg. Co.	Shyrock	New Home
Sedwig, J.	A. G. Mason Mfg. Co.	Shyrock	White
Sefton	A. G. Mason Mfg. Co.	Shyrock, The	A. G. Mason Mfg. Co.
Segal Special	National	Shuff's Own	Standard
Segnor	Standard	Shurley	A. G. Mason Mfg. Co.

Sewing-Machine Name	Sewing-Machine Company	Sewing-Machine Name	Sewing-Machine Company
Sidders	Free	Smytha	White
Siebold Special	A. G. Mason Mfg. Co.	Smythe and Co.	National
Siegel	Free	Snodgrass	A. G. Mason Mfg. Co.
Siegfried	White	Snow	A. G. Mason Mfg. Co.
Sierers	White	Snavely	National
Sigman	A. G. Mason Mfg. Co.	Snellenburg	New Home
Signal	Davis	Snyder	A. G. Mason Mfg. Co.
Signal	Standard	Soden	Davis
Signor	Goodrich	So Easy	Free
Silco	Free	Soeasy	National
Silent	Davis	So-Ezy	Davis
Silent Princess	National	So- Ezy	New Home
Silent Stitcher	Davis	Solomons	Standard
Silvens	A. G. Mason Mfg. Co.	Solon	A. G. Mason Mfg. Co.
Silver	Free	Sonora	Standard
Silver King	Davis	Sonoria	National
Silver Medal	National	Soo Special	National
Silver Spray	Davis	Sorosis	National
Silver Star A, B, C, D, II	Davis	Souder	White
Silveus	White	South Bend	White
Simmons Co.	National	Southern	Davis
Simon	A. G. Mason Mfg. Co.	Southern	White
Simplex	Davis	Southern Belle	Davis
Simplex	Free	Southern Novelty	Free
Simplex	White	Southwell	National
Simpson and Will	A. G. Mason Mfg. Co.	Sowerset	White
Sine, H. L.	A. G. Mason Mfg. Co.	Spalding	Standard
Singleton	White	Spartan	Free
Sin Par	White	Spear Edge	National
Sinsel	White	Special	A. G. Mason Mfg. Co.
Sit-Easy	White	Special	Standard
Sitright Franklin, The	Domestic	Spee Dee	Standard
Sit-Straight	White	Speedway	National
Skandanavian	Davis	Speedwell	National
Skeens	White	Spell, F. C.	A. G. Mason Mfg. Co.
Skeleton	National	Spencer	Davis
Skinner's	White	Spencer	A. G. Mason Mfg. Co.
Skitler	A. G. Mason Mfg. Co	Spetnagel	Davis
Slane, The	National	Spicer	A. G. Mason Mfg. Co.
Slater	Standard	Spike Nash Co.	A. G. Mason Mfg. Co.
Slaughter	White	Spill Brothers	White
Slavic	National	Spindall	A. G. Mason Mfg. Co.
Sleepine	White	Spitler	White
Slidell	Standard	Splendid	Free
Sluder	White	Spokane	New Home
Smalley	National	Spotless	National
Small, M.	White	Sprague	A. G. Mason Mfg. Co.
Smathers	White	Spring City	Standard
Smith	Davis	Spring City	White
Smith	White	Spurlin	White
Smith Special	New Home	Square Deal	A. G. Mason Mfg. Co.
Smith and Caugheys	Standard	Srojan	National
Smith and Wesson	Free	Srormost	Goodrich

Sewing-Machine Name	Sewing-Machine Company	Sewing-Machine Name	Sewing-Machine Company
Stag	Davis	Stimulator	New Home
Stahns Favorite	Standard	Stitchwell	National
Stahn's, E. A.	A. G. Mason Mfg. Co.	Stitchwell	Standard
Standart	Davis	Stockdale	A. G. Mason Mfg. Co.
Stand, New	A. G. Mason Mfg. Co.	Stockdale Rotary	White
Standard (after 27500)	Standard	Stockman	A. G. Mason Mfg. Co.
Standard Electric	Standard	Stockman & Farmer	A. G. Mason Mfg. Co.
Standard Favorite	Standard	Stone	A. G. Mason Mfg. Co.
Standard, J. W.	A. G. Mason Mfg. Co.	Stoops	Standard
Standard Norwood	Standard	Store Big Special	Goodrich
Standard Paragon	Standard	Stores Grenada	National
Standard Rotary	Standard	Stork	New Home
Standard Vibrator	Standard	Stouffer	White
Standfield, J. W.	A. G. Mason Mfg. Co.	Stoval	A. G. Mason Mfg. Co.
Stanffer	White	Strahn's Favorite	Standard
Stanley	National	Straightway	Standard
Stanly	Standard	Strange	National
Stapp	A. G. Mason Mfg. Co.	Strangburg	A. G. Mason Mfg. Co.
Star	Davis	Strand's Electric	A. G. Mason Mfg. Co.
Star	A. G. Mason Mfg. Co.	Straco	Free
Star Capoc	A. G. Mason Mfg. Co.	Stratton	White
Star Golden	A. G. Mason Mfg. Co.	Strawbridge Clothier	National
Star Installment	A. G. Mason Mfg. Co.	Street	A. G. Mason Mfg. Co.
Star New Home	New Home	Streets	White
Staufer	Davis	Strelezpk	A. G. Mason Mfg. Co.
Stee Merc. Co.	A. G. Mason Mfg. Co.	Strong Rotary Electric	White
Steel's Favorite	A. G. Mason Mfg. Co.	Strongs Favorite	A. G. Mason Mfg. Co.
Steengagen	Standard	Stronds Electric	A. G. Mason Mfg. Co.
Stager	National	Stroud Rotary Electric	White
Stegman Electric Rotary	White	Struthers Best	Standard
Stein	Davis	Stubbo	White
Stein	A. G. Mason Mfg. Co.	Stulco	White
Stein Furniture Company	Standard	Style Mode	Davis
Steinkamp	White	Success	Free
Stella	National	Successful	National
Stemkamp	Standard	Sullivan	Davis
Stemway	National	Sullivan Slade Co.	A. G. Mason Mfg. Co.
Stephenson	A. G. Mason Mfg. Co.	Sultan	Davis
Stepp, J. M.	A. G. Mason Mfg. Co.	Sultana	National
Sterch	A. G. Mason Mfg. Co.	Summers	A. G. Mason Mfg. Co.
Sterling	Davis	Summit	A. G. Mason Mfg. Co.
Sterling	Standard	Sun and Voice	Davis
Sterling A, B, C, D, E, F	Goodrich	Sunbeam	White
Stern	Davis	Sunflower	National
Sternberg	A. G. Mason Mfg. Co.	Suniers	White
Sternikamp	A. G. Mason Mfg. Co.	Sunoria	Free
Stevens	National	Sunray	Davis
Stewart	A. G. Mason Mfg. Co.	Sunset	Davis
Stewart Special	Free	Sunset	A. G. Mason Mfg. Co.
Stier Special	A. G. Mason Mfg. Co.	Sunshine	Davis
Stigall and Potts	A. G. Mason Mfg. Co.	Sunshine	Goodrich
Stilles	White	Sunney South	A. G. Mason Mfg. Co.
Stilleto	Free	Superb	Davis

Sewing-Machine Name	Sewing-Machine Company	Sewing-Machine Name	Sewing-Machine Company
Superb	White	Therlaigs	A. G. Mason Mfg. Co.
Superba	Davis	Thelma	Standard
Superior	Davis	Thelma's Favorite	A. G. Mason Mfg. Co.
Superior	New Home	Thistle	National
Superior	Standard	Thomas	Davis
Super Quality	Standard	Thompson	A. G. Mason Mfg. Co.
Supplies 2	Davis	Thompson	Standard
Supreme	Davis	Tidente	New Home
Supreme	Goodrich	The 400	Standard
Surasky Special	Free	The Tiger	Davis
Surelock Rotary	National	Ti-Ki	Goodrich
Surety	Davis	Tillsworth	A. G. Mason Mfg. Co.
Surprise	Standard	Tisbest	A. G. Mason Mfg. Co.
Surprise	A. G. Mason Mfg. Co.	Tissier Special	A. G. Mason Mfg. Co.
Surviver, J. S.	A. G. Mason Mfg. Co.	Tobin Bros.	A. G. Mason Mfg. Co.
Sutton Bros.	A. G. Mason Mfg. Co.	Todd, H. E.	A. G. Mason Mfg. Co.
Swan	A. G. Mason Mfg. Co.	Toledo	Goodrich
Swan	Goodrich	Tomanis	National
Swank	Davis	Tom Cummins	National
Swastika	Standard	Tomec, The	A. G. Mason Mfg. Co.
Sweeton and B.	White	Tondy	Standard
Swift	Standard	Townley	Davis
Swinton	Standard	Trabue	A. G. Mason Mfg. Co.
Suwanee, A. C.	Free	Treasure	Goodrich
Sweden	A. G. Mason Mfg. Co.	Treasurer	A. G. Mason Mfg. Co.
Symonds	Standard	Tremont	Davis
Syndicate	National	Trennan	A. G. Mason Mfg. Co.
Syracuse	National	Trenton	New Home
		Trice	A. G. Mason Mfg. Co.
Tabin	White	Trice Bros.	National
Taff	Goodrich	Tri-City	National
Tallulah, The	A. G. Mason Mfg. Co.	Trinacria	Standard
Tanek, M.	A. G. Mason Mfg. Co.	Trip Imp.	National
Target	Standard	Tristate	Davis
Tarnocks	A. G. Mason Mfg. Co.	Triumph	Davis
Tarver	A. G. Mason Mfg. Co.	Triumph	New Home
Tates	A. G. Mason Mfg. Co.	Trotter	National
Taylor	Davis	Tuapecka	Standard
Taylor	A. G. Mason Mfg. Co.	Tucker	National
Taylor	Standard	Tuckwell	Standard
Teague	Davis	Tude 7A	Standard
Tecco	Standard	Tuggers	A. G. Mason Mfg. Co.
Tedstorm	White	Tuggle, H. D.	A. G. Mason Mfg. Co.
Tedstrom	A. G. Mason Mfg. Co.	Tupelo Mer. Co.	National
Tellico	National	Turn City	National
Temple	National	Turners	A. G. Mason Mfg. Co.
Tenks	Davis	Turnocks	A. G. Mason Mfg. Co.
Tennessee	A. G. Mason Mfg. Co.	Turpin's Pride	A. G. Mason Mfg. Co.
Tenn Furniture Co.	A. G. Mason Mfg. Co.	Tuscarawas	A. G. Mason Mfg. Co.
Terrebone	A. G. Mason Mfg. Co.	Tuxedo	New Home
Teutonia	New Home	Tuxhorns	A. G. Mason Mfg. Co.
Texas	A. G. Mason Mfg. Co.	Twentieth Century	A. G. Mason Mfg. Co.
Texas	White	Twews	White

Sewing-Machine Name	Sewing-Machine Company	Sewing-Machine Name	Sewing-Machine Company
Two-in-One	Standard	Vaughan	Davis
Tyanko	National	Vaughan	Standard
Tyler	A. G. Mason Mfg. Co.	Velox	National
Tyrholm	A. G. Mason Mfg. Co.	Velox 2	Davis
		Velox A & B	New Home
Ulasta	Goodrich	Venango	Goodrich
Ulmer	A. G. Mason Mfg. Co.	Vendome	National
Unaka	Free	Venture	Davis
Unanzst	Standard	Venus	National
U-Need-Me	A. G. Mason Mfg. Co.	Verhart, F. M.	Standard
Uni	A. G. Mason Mfg. Co.	Veribest	Standard
Unica	A. G. Mason Mfg. Co.	Vertical Feed	Davis
Unicorn	A. G. Mason Mfg. Co.	Via and Stedman	A. G. Mason Mfg. Co.
Union	Davis	Vibra	Davis
Union	New Home	Vibrant	Standard
Union 24	Standard	Vibrator	Standard
Union Advance	Davis	Vibratorio	A. G. Mason Mfg. Co.
Union American	Davis	Viceroy	National
Union Grand	Free	Vicellio	A. G. Mason Mfg. Co.
Union Leader	National	Victor	Davis
Union Merc. Co.	A. G. Mason Mfg. Co.	Victor, New	A. G. Mason Mfg. Co.
Unique	Davis	Victor 03	Free
Unique	A. G. Mason Mfg. Co.	Victoria	Standard
United	National	Victoria	A. G. Mason Mfg. Co.
United States	A. G. Mason Mfg. Co.	Victoria Caniff	National
Unito	National	Victorio	Davis
Unity	A. G. Mason Mfg. Co.	Victory	Davis
Universal	Davis	Victory	Free
Universal	New Home	Victory, New, 9	Goodrich
Universal	Standard	Vigo	Davis
Up-To-Date	Davis	Viking	Davis
Up-To-Date	New Home	Viking	A. G. Mason Mfg. Co.
Upright	Standard	Vini Model	National
U.S. Dailey	Standard	Vindex	National
Utah	Free	Vinegard	A. G. Mason Mfg. Co.
Utica	Davis	Virginia	National
Utility	Davis	Virginian	National
Utility	Standard	Vittorio, E. M.	A. G. Mason Mfg. Co.
Utopia	Goodrich	Voekel	A. G. Mason Mfg. Co.
		Vogel	A. G. Mason Mfg. Co.
Vallarena	National	Vogelsburg	A. G. Mason Mfg. Co.
Valley City	Free	Vogue	Free
Van Allen	A. G. Mason Mfg. Co.	Vogue	A. G. Mason Mfg. Co.
Van Loon Special	Free	Vola	National
Vance A	National	Volksfriend	National
Vance, The	Davis	Volkspost	National
Vancello	National	Volo	Davis
Vanderbilt	Davis	Volo	National
Vanters	Goodrich	Volunteer	Davis
Varnado	Standard	Volunteer	A. G. Mason Mfg. Co.
Vassar	National	Vords	National
Vassar	Standard	Vose, V. S.	A. G. Mason Mfg. Co.
Vasumpaur	Free	Voucher	Standard

Sewing-Machine Name	Sewing-Machine Company	Sewing-Machine Name	Sewing-Machine Company
Vulcan	National	Warren	Standard
Vulcan	Standard	Wartanberg	A. G. Mason Mfg. Co.
		Warwick, New	A. G. Mason Mfg. Co.
Wabash	Davis	Wasans	A. G. Mason Mfg. Co.
Wabash	New Home	Washington	New Home
Wabash	Standard	Washington	Standard
Wa-Car-Co	A. G. Mason Mfg. Co.	Washtenaw	National
Wacerly	National	Watango	Goodrich
Wade	A. G. Mason Mfg. Co.	Watchman	National
Wagener	Goodrich	Watkins	A. G. Mason Mfg. Co.
Wagley	A. G. Mason Mfg. Co.	Watson	National
Wagner H	Free	Wauson	National
Wah Wah	National	Waverly	A. G. Mason Mfg. Co.
Wako	National	Waxahacie Hardware Co.	A. G. Mason Mfg. Co.
Wainright	Standard	Wayne	National
Walbert	Goodrich	Waywick New	Free
Walbridge	Davis	Weal	National
Walcot	A. G. Mason Mfg. Co.	Wearwell	National
Waldo	Goodrich	Weatherford	A. G. Mason Mfg. Co.
Waldorf	Davis	Weathers	White
Waldorf	A. G. Mason Mfg. Co.	Weatherspoon	A. G. Mason Mfg. Co.
Walker	Davis	Weaver	Goodrich
Walker	National	Weaver	White
Wallace	Free	Webb	A. G. Mason Mfg. Co.
Wallavok	Goodrich	Weber	Standard
Walters	A. G. Mason Mfg. Co.	Weber	White
Waltham	Davis	Weber Special	Free
Waltham	Free	Weels	White
Waltham Rotary	National	Weerooms	White
Walton	A. G. Mason Mfg. Co.	Weesmer	White
Wanamaker Rotary	White	Weesner Ream Co.	Standard
Wanamaker, S. R.	White	Weeter	Davis
Wanda	Goodrich	Weichers	A. G. Mason Mfg. Co.
Wanemaker	New Home	Weigand	A. G. Mason Mfg. Co.
Wankon	White	Weig and Schultz	Standard
Wanless	Goodrich	We'ils	White
Wannamaker	Davis	Weiners	White
Wannamaker	New Home	Weirs	A. G. Mason Mfg. Co.
Wanzer C	Standard	Weisberger	Davis
Wapello Chief	A. G. Mason Mfg. Co.	Weise	Davis
Warbler	Free	Welborn Bros.	Standard
Ward	Unknown	Welch	White
Ward	Standard	Welcome	A. G. Mason Mfg. Co.
Ward, New	National	Welcome	Standard
Wardell	A. G. Mason Mfg. Co.	Weldin	Standard
Wardlan	A. G. Mason Mfg. Co.	Wellington	Davis
Ware	A. G. Mason Mfg. Co.	Wellington	A. G. Mason Mfg. Co.
Warewell A, C, E, G, H	National	Wells	A. G. Mason Mfg. Co.
Warkatz	Standard	Wellworth	Free
Warnell	A. G. Mason Mfg. Co.	Wellworth E.	National
Warner, The	A. G. Mason Mfg. Co.	Welsh John	A. G. Mason Mfg. Co.
Warnocks	A. G. Mason Mfg. Co.	Wentworth	National
Warren	A. G. Mason Mfg. Co.	Werner	Davis

Sewing-Machine Name	Sewing-Machine Company	Sewing-Machine Name	Sewing-Machine Company
Werner	A. G. Mason Mfg. Co.	Wilkins and Field	Standard
Werner, The	Standard	Will C. Free	Free
Wertenberger	A. G. Mason Mfg. Co.	Willamette	National
Wertz, W. D.	A. G. Mason Mfg. Co.	Willard	Davis
West	White	Willford	National
West Brothers	A. G. Mason Mfg. Co.	William Penn	Free
Western	A. G. Mason Mfg. Co.	Williams	National
Western Electric	National	Williams	White
Western Queen	A. G. Mason Mfg. Co.	Williams, New	Unknown
Western Union	A. G. Mason Mfg. Co.	Williams Singer Model	Unknown
Westlake	National	Williamson	A. G. Mason Mfg. Co.
Westinghouse	National	Willington	A. G. Mason Mfg. Co.
Westmoreland	National	Willis	Standard
Westrailia	A. G. Mason Mfg. Co.	Willis and Co.	A. G. Mason Mfg. Co.
Wetherbee	White	Willis, J. M. T. A.	Standard
Wetherbee, H., and Co.	A. G. Mason Mfg. Co.	Wilmot	National
Weyand Bros.	A. G. Mason Mfg. Co.	Wilson	A. G. Mason Mfg. Co.
Weyth	National	Wilson Rotary	A. G. Mason Mfg. Co.
Wheeleer	Free	Will Ranciers	A. G. Mason Mfg. Co.
Wheeler	White	Winchester	Davis
Wheeler A, B, C, E, P, W	New Home	Winchester	Goodrich
Wheeling	Davis	Winchester	Free
White, A. J. Ltd.	Standard	Winchester	White
White-Cross	Davis	Wingold	Goodrich
White Diamond	Standard	Winn	A. G. Mason Mfg. Co.
White Family Rotary	White	Winnepag	Free
White Family Shuttle	White	Winnepeg	Free
		Winner	Davis
White Furniture Co.	A. G. Mason Mfg. Co.	Winner	Free
White House	Davis	Winner, New	A. G. Mason Mfg. Co.
White King	Standard	Winslow Special	A. G. Mason Mfg. Co.
White New Family Sewing Machine	White	Wise, The	A. G. Mason Mfg. Co.
		Wise-Rotary, The	White
White Peerless	A. G. Mason Mfg. Co.	Witherbee	A. G. Mason Mfg. Co.
White Pet	A. G. Mason Mfg. Co.	Witherow	Goodrich
White Pet	White	Witners	National
White Star	Free	Wittes Rotary	National
White Vibrator	White	Wizard	Davis
Whitehall, New	Free	Wizard	Standard
Whitehall, Old	Unknown	Wolfe	A. G. Mason Mfg. Co.
Whitehead	National	Wolters	A. G. Mason Mfg. Co.
Whitehead	White	Wolverine	Free
Whitehouse	A. G. Mason Mfg. Co.	Wolverine	A. G. Mason Mfg. Co.
Whitelau	A. G. Mason Mfg. Co.	Woman and Home	New Home
Whitely	A. G. Mason Mfg. Co.	Woman & House	New Home
Whitemarsh	White	Women's Institute Rotary	White
Whiteside	A. G. Mason Mfg. Co.	Wonder	Davis
Whiting	Standard	Wonder	Free
Whitner & R	A. G. Mason Mfg. Co.	Wonder, The	A. G. Mason Mfg. Co.
Whitsett	A. G. Mason Mfg. Co.	Wonder Worker	Standard
Wigginton	A. G. Mason Mfg. Co.	Wood	A. G. Mason Mfg. Co.
Wild Rose	National	Woodbert	A. G. Mason Mfg. Co.
Wilks	A. G. Mason Mfg. Co.	Woodhill	Standard

Sewing-Machine Name	Sewing-Machine Company	Sewing-Machine Name	Sewing-Machine Company
Woodidle	White	Xmas	Standard
Woodman	New Home		
Wood's Special	A. G. Mason Mfg. Co.	Yahrborough	White
Woodrill	A. G. Mason Mfg. Co.	Yale	Davis
Woodruff	Davis	Yale	Free
Woodward	Davis	Yankee	Standard
Woodworth	Standard	Yarborough	A. G. Mason Mfg. Co.
Woolbert	A. G. Mason Mfg. Co.	Yates	White
Woolsey	A. G. Mason Mfg. Co.	Yews	National
Wootherspoon	Standard	Yoder and McLean	A. G. Mason Mfg. Co.
Workwell	Standard	York	A. G. Mason Mfg. Co.
World Rotary	A. G. Mason Mfg. Co.	York Music Co.	A. G. Mason Mfg. Co.
Worlds Best	Davis	York and Wadsworth	A. G. Mason Mfg. Co.
Worlds Fair	National	Young Blood	A. G. Mason Mfg. Co.
Worlds Fair Premium	A. G. Mason Mfg. Co.	Young Brothers	Standard
Worlds Rotary	White	Young's Favorite	Standard
		Young's Reliable	A. G. Mason Mfg. Co.
Wormack, R. V. A.	A. G. Mason Mfg. Co.	Young and Chaffee	National
Wornell	White	Young and Maritin	A. G. Mason Mfg. Co.
Worthmore, E	Free	Young and McComb	National
Worthmore, E	National	Youman and Leete	A. G. Mason Mfg. Co.
Wright	A. G. Mason Mfg. Co.	Youth's Companion	New Home
Wrights	Davis	Yvins Leader	Goodrich
Wrights	A. G. Mason Mfg. Co.		
Wrigley	Free	Zachman	A. G. Mason Mfg. Co.
Wrinkle, M. Co.	A. G. Mason Mfg. Co.	Zawadski	White
Wu Yeo Co.	Standard	Zenith	National
		Zenith A	Free
Wyandott Furniture Co.	A. G. Mason Mfg. Co.	Zenobla	A. G. Mason Mfg. Co.
Wyatt	White	Zephyr Mer. Co.	National
Wyeth	National	Ziberna	White
Wymes	White	Zinco	National
Wyness	White	Zinco City	National
Wyoming	Standard	Zuberma	National
		Zulema	National
Xmas	Goodrich		

V. Chronological List of U.S. Sewing-Machine Patent Models in the Smithsonian Collections

There are more than seven hundred sewing-machine patent models and a similar number of attachment models in the Smithsonian collections. Most of these machines were received in 1926 when the Patent Office disposed of its collection of hundreds of thousands of models. Prior to 1880, models had been required with the patent application; although the requirement was discontinued that year, patentees continued to furnish models for another decade or so. All models prior to 1836 were lost in a Patent Office fire of that year, but since the sewing-machine patent history dates from the 1840s, most of the historically important ones of this subject have been preserved.

These models form a valuable part of the record of the invention, supplementing the drawings and the text of the written specifications. The early sewing-machine models were made to order, either by the inventor or a commissioned model maker. As soon as sewing machines were produced commercially, it was less expensive for the patentee to use a commercial machine of the period, to which he added his change or improvement, than to have a complete model constructed to order. Some of the commercial machines used in this way are the only examples known to be in existence and, as such, are of more interest in establishing the history of the manufactured machine than for the minor patented changes.

During the period of the "Sewing-Machine Combination," many patentees attempted to invent and patent "the different machine." This was either a radical change in style or an attempt to produce a far less-expensive type of machine. These machines were not always put into commercial production, but the patent models give an indication of the extent to which some inventors went to simplify or vary the mechanics of machine sewing.

The following is a list of those sewing-machine patent models in the Smithsonian Institution collections:

Patentee	Date	Patent Number
Greenough, John J.	Feb. 21, 1842	2,466
Bean, Benjamin W.	March 4, 1843	2,982
Corliss, George H.	Dec. 27, 1843	3,389
Howe, Elias, Jr.	Sept. 10, 1846	4,750
Bachelder, John	May 8, 1849	6,439
Wilson, Allen B.	Nov. 12, 1850	7,776
Robinson, Frederick R.	Dec. 10, 1850	7,824
Grover & Baker	Feb. 11, 1851	7,931
Singer, Isaac M.	Aug. 12, 1851	8,294
Wilson, Allen B.	Aug. 12, 1851	8,296
Wilson, Allen B.	June 15, 1852	9,041
Miller, Charles	July 20, 1852	9,139
Avery, Otis	Oct. 19, 1852	9,338
Hodgkins, C.	Nov. 2, 1852	9,365
Bradeen, J. G.	Nov. 2, 1852	9,380
Bates, W. G.	Feb. 22, 1853	9,592
Thompson, T. C.	March 29, 1853	9,641
Wickersham, W.	April 19, 1853	9,679
Johnson, W. H.	March 7, 1854	10,597
Harrison, J., Jr.	April 11, 1854	10,763
Avery, Otis	May 9, 1854	10,880
Singer, Isaac	May 30, 1854	10,975
Hunt, Walter	June 27, 1854	11,161
Roper, S. H.	Aug. 15, 1854	11,531
Shaw, P.	Sept. 12, 1854	11,680
Ambler, D. C.	Nov. 1, 1854	11,884
Robertson, T. J. W.	Nov. 28, 1854	12,015
Lyon, W.	Dec. 12, 1854	12,066
Stedman, G. W.	Dec. 12, 1854	12,074
Ward, D. T.	Jan. 2, 1855	12,146
Conant, J. S.	Jan. 16, 1855	12,233
Smith, H. B.	Jan. 16, 1855	12,247
Singer, I. M.	Feb. 6, 1855	12,364
Stedman, G. W.	March 20, 1855	12,573
Stedman, G. W.	May 1, 1855	12,798

Patentee	Date	Patent Number
Chilcott, J., and Scrimgeour, J.	March 15, 1855	12,856
Durgin, Charles A.	May 22, 1855	12,902
Bond, J., Jr.	May 22, 1855	12,939
Singer, Isaac	June 12, 1855	13,065
Harrison, J., Jr.	Oct. 2, 1855	13,616
Singer, I. M.	Oct. 9, 1855	13,661
Singer, I. M.	Oct. 9, 1855	13,662
Langdon, L. W.	Oct. 30, 1855	13,727
Stedman, G. W.	Nov. 27, 1855	13,856
Swingle, A.	Feb. 5, 1856	14,207
Watson, Wm. C.	March 11, 1856	14,433
Singer, I. M.	March 18, 1856	14,475
Grover, W. O.	May 27, 1856	14,956
Blodgett, S. C.	Aug. 5, 1856	15,469
Roper, S. H.	Nov. 4, 1856	16,026
Singer, Isaac M.	Nov. 4, 1856	16,030
Gibbs, James E. A.	Dec. 16, 1856	16,234
Jennings, L.	Dec. 16, 1856	16,237
Johnson, A. F.	Jan. 13, 1857	16,387
Gibbs, J. E. A.	Jan. 20, 1857	16,434
Howe, Elias, Jr.	Jan. 20, 1857	16,436
Alexander, Elisa	Feb. 3, 1857	16,518
Gray, Joshua	Feb. 3, 1857	16,566
Belcher, C. D.	March 3, 1857	16,710
Pratt, S. F.	March 3, 1857	16,745
Nettleton & Raymond	April 14, 1857	17,049
Gibbs, J. E. A.	June 2, 1857	17,427
Harris, Daniel	June 9, 1857	17,508
Harris, Daniel	June 16, 1857	17,571
Sage, William	June 30, 1857	17,717
Lathbury, E. T.	July 7, 1857	17,744
Wickersham, W.	Aug. 25, 1857	18,068
Wickersham, W.	Aug. 25, 1857	18,069
Behn, Henry	Aug. 25, 1857	18,071
Nettleton, Wm. H., and Raymond, Charles	Oct. 6, 1857	18,350
Roper, S. H.	Oct. 27, 1857	18,522
Fetter, George	Dec. 1, 1857	18,793
Watson, W. C.	Dec. 8, 1857	18,834
Behn, H.	Dec. 15, 1857	18,880
Hubbard, George W.	Dec. 22, 1857	18,904
Lazelle, W. H.	Dec. 22, 1857	18,915
Clark, David W.	Jan. 5, 1858	19,015
Fetter, George	Jan. 5, 1858	19,059
Clark, David W.	Jan. 12, 1858	19,072
Clark, David W.	Jan. 19, 1858	19,129
Dimmock, Martial, and Rixford, Nathan	Jan. 19, 1858	19,135
Boyd, A. H.	Jan. 19, 1858	19,171
Angell, Benjamin J.	Feb. 9, 1858	19,285
Clark, David W.	Feb. 23, 1858	19,409
Raymond, Charles	March 9, 1858	19,612
Hendrick, Joseph E.	March 16, 1858	19,660
Parker, Sidney	March 16, 1858	19,662
Gray, Joshua	March 16, 1858	19,665
Coates, F. S.	March 23, 1858	19,684
Clark, David W.	March 23, 1858	19,732
Reynolds, O. S.	March 30, 1858	19,793
Bartholf, Abraham	April 6, 1858	19,823
Savage, E.	April 6, 1858	19,876
Atwood, J. E., J. C., and O.	April 13, 1858	19,903
Bosworth, Chas. F.	April 20, 1858	19,979
Clark, David W.	June 8, 1858	20,481
Herron, A. C.	June 15, 1858	20,557
Johnson, A. F.	June 22, 1858	20,686
Barnes, W. T.	June 29, 1858	20,688
Smith, E. H.	June 29, 1858	20,739
West, H. B., and Willson, H. F.	June 29, 1858	20,753
Miller, W.	June 29, 1858	20,763
Blake, Lyman R.	July 6, 1858	20,775
Carpenter, Lunan	July 27, 1858	20,990
Moore, Charles	July 27, 1858	21,015
Smith, E. H.	Aug. 3, 1858	21,089
Wheeler and Carpenter	Aug. 3, 1858	21,100
Gibbs, J. E. A.	Aug. 10, 1858	21,129
Uhlinger, W. P.	Aug. 17, 1858	21,224
Clark, David W.	Aug. 31, 1858	21,322
Blodgett, S. C.	Sept. 7, 1858	21,465
Hubbard, G. W.	Sept. 14, 1858	21,537
Hendrick, J. E.	Oct. 5, 1858	21,722
Gibbs, J. E. A.	Oct. 12, 1858	21,751
Sangster, Amos. W.	Oct. 26, 1858	21,929
Avery, O. and Z. W.	Nov. 9, 1858	22,007
Spencer and Lamb	Nov. 23, 1858	22,137
Perry, James	Nov. 23, 1858	22,148
Burnet and Broderick	Nov. 30, 1858	22,160
Hook, Albert H.	Nov. 30, 1858	22,179
Raymond, Charles	Nov. 30, 1858	22,220
Bishop, H. H.	Dec. 7, 1858	22,226
Pratt, S. F.	Dec. 7, 1858	22,240
Atwood, J. E.	Dec. 14, 1858	22,273
Fosket, W. A., and Savage, Elliot	Jan. 25, 1859	22,719
Snyder, W.	Feb. 15, 1859	22,987
Clark, D. W.	May 3, 1859	23,823
Boyd, A. H.	May 17, 1859	24,003
Gray, Joshua	May 17, 1859	24,022
Hook, Albert H.	May 17, 1859	24,027
Spencer, James C.	May 17, 1859	24,061
Carhart, Peter S.	May 24, 1859	24,098
McCurdy, J. S.	June 14, 1859	24,395

Patentee	Date	Patent Number	Patentee	Date	Patent Number
Goodwyn, H. H.	June 21, 1859	24,455	Gibbs, J. E. A.	June 26, 1860	28,851
Grout, William	July 5, 1859	24,629	McCurdy, J. S.	July 3, 1860	28,993
Hensel, George	July 12, 1859	24,737	Mueller, H.	July 3, 1860	28,996
Parker, Sidney	July 12, 1859	24,780	Sutton, Wm. A.	July 17, 1860	29,202
Hall, William	July 26, 1859	24,870	Hicks, W. C.	July 24, 1860	29,268
Hayden, H. W.	Aug. 2, 1859	24,937	Tracy, D.	Sept. 11, 1860	30,012
Kelsey, D.	Aug. 2, 1859	24,939	Washburn, T. S.	Sept. 11, 1860	30,031
Emswiler, J. B.	Aug. 9, 1859	25,002	Arnold, G. B., and A.	Sept. 25, 1860	30,112
Farr, C. N.	Aug. 9, 1859	25,004	Leavitt, Rufus	Nov. 13, 1860	30,634
Harrison, James, Jr.	Aug. 9, 1859	25,013	Payne, R. S.	Nov. 13, 1860	30,641
Tapley, G. S.	Aug. 9, 1859	25,059	Heyer, Frederick	Nov. 27, 1860	30,731
Barnes, W. T.	Aug. 16, 1859	25,084	Hardie, J. W.	Dec. 4, 1860	30,854
Booth, Ezekial	Aug. 16, 1859	25,087	Earle, T.	Jan. 22, 1861	31,156
Hinkley, J.	Aug. 23, 1859	25,231	Bruen, J. T.	Jan. 22, 1861	31,208
Harrison, James, Jr.	Aug. 30, 1859	25,262	Smith, J. M.	Feb. 5, 1861	31,334
Buell, J. S.	Sept. 13, 1859	25,381	Smith, L. H.	Feb. 12, 1861	31,411
Vogel, Kasimir	Oct. 4, 1859	25,692	Rice, Quartus	Feb. 12, 1861	31,429
Woodward, F. G.	Oct. 11, 1859	25,782	Rose, I. M.	March 5, 1861	31,628
Barrett, O. D.	Oct. 11, 1859	25,785	Ross, Noble G.	March 26, 1861	31,829
Barnes, William T.	Oct. 25, 1859	25,876	Boyd, A. H.	April 2, 1861	31,864
Sawyer, Irwin, and Alsop, T.	Oct. 25, 1859	25,918	Mallary, G. H.	April 2, 1861	31,897
Budlong, William G.	Nov. 1, 1859	25,946	Shaw, H. L.	April 9, 1861	32,007
Fosket, William A., and Savage, E.	Nov. 1, 1859	25,963	Burr, Theodore	April 9, 1861	32,023
			Jones, William, and Haughian, P.	May 14, 1861	32,297
Hicks, W. C.	Nov. 8, 1859	26,035			
Scofield, C.	Nov. 8, 1859	26,059	Wilder, M. G.	May 14, 1861	32,323
Pearson, William	Nov. 22, 1859	26,201	Smith, Lewis H.	May 21, 1861	32,385
McCurdy, James S.	Nov. 22, 1859	26,234	Stoakes, J. W.	May 28, 1861	32,456
Clark, Edwin	Dec. 6, 1859	26,336	Fuller, William M.	June 4, 1861	32,496
Dickinson, C. W.	Dec. 6, 1859	26,346	Norton, B. F.	July 9, 1861	32,782
Miller, Charles	Dec. 13, 1859	26,462	Raymond, C.	July 9, 1861	32,785
Rowe, Jas.	Dec. 27, 1859	26,638	Raymond, Charles	July 30, 1861	32,925
Johnson, A. F.	Jan. 24, 1860	26,948	Case, G. F.	Aug. 13, 1861	33,029
Thomson, J.	Feb. 7, 1860	27,082	Hodgkins, C.	Aug. 20, 1861	33,085
Juengst, George	Feb. 14, 1860	27,132	Marble, F. E.	Oct. 8, 1861	33,439
Davis, Job A.	Feb. 21, 1860	27,208	Mann, Charles	Oct. 22, 1861	33,556
Gibbs, James E. A.	Feb. 21, 1860	27,214	Grover, W. O.	Nov. 26, 1861	33,778
Rowe, James	Feb. 21, 1860	27,260	Hendrickson, E. M.	Feb. 4, 1862	34,330
Dopp, H. W.	Feb. 28, 1860	27,279	Derocquigny, A. C. F., Gance, D., and Hanzo, L.	March 25, 1862	34,748
Paine, A. R.	March 6, 1860	27,412			
Smalley, J.	March 20, 1860	27,577	Thompson, R.	April 8, 1862	34,926
Newlove, T.	April 3, 1860	27,761	Smith, John C.	April 15, 1862	34,988
McCurdy, J. S.	May 1, 1860	28,097	Palmer, Aaron	May 13, 1862	35,252
Arnold, G. B.	May 8, 1860	28,139	Hall, W. S.	Aug. 5, 1862	36,084
Bean, E. E.	May 8, 1860	28,144	McCurdy, James S.	Aug. 19, 1862	36,256
Holly, Birdsill	May 8, 1860	28,176	Grover, W. O.	Sept. 9, 1862	36,405
Chamberlain, J. N.	May 29, 1860	28,452	Wilkins, J. N.	Sept. 30, 1862	36,591
Ruddick, H.	May 29, 1860	28,538	Humphrey, D. W. G.	Oct. 7, 1862	36,617
Scofield, Chas., and Rice, Clarke	June 5, 1860	28,610	House, H. A., and J. A.	Nov. 11, 1862	36,932
			Crossby, C. O., and Kellogg, H.	Dec. 2, 1862	37,033
Smith, Wilson H.	June 19, 1860	28,785			
Rose, I. M.	June 19, 1860	28,814	Shaw, A. B.	Dec. 16, 1862	37,202

Patentee	Date	Patent Number	Patentee	Date	Patent Number
Pipo, John A.	Jan. 27, 1863	37,550	Rehfuss, George	Nov. 21, 1865	51,086
Hollowell, J. G.	Feb. 10, 1863	37,624	Eickemeyer, Rudolf	Feb. 20, 1866	52,698
Howe, A. B.	March 17, 1863	37,913	Hanlon, John	Feb. 27, 1866	52,847
Weitling, W.	March 17, 1863	37,931	McCurdy, J. S.	April 3, 1866	53,743
Shaw & Clark	April 21, 1863	38,246	Bartram, W. B.	May 15, 1866	54,670
Baldwin, Cyrus W.	April 28, 1863	38,276	Bartram, W. B.	May 15, 1866	54,671
Grote, F. W.	May 5, 1863	38,447	Goodspeed, G. N.	May 15, 1866	54,816
Palmer, C. H.	May 5, 1863	38,450	Hayes, J.	May 22, 1866	55,029
Mack, W. A.	May 19, 1863	38,592	McCloskey, John	June 19, 1866	55,688
Bosworth, C. F.	June 9, 1863	38,807	House, J. A. and H. A.	June 26, 1866	55,865
McCurdy, J. S.	June 16, 1863	38,931	Tucker, Joseph C.	July 24, 1866	56,641
Langdon, Leander W.	July 14, 1863	39,256	Warth, Albin	July 24, 1866	56,646
House, J. A., and H.A.	Aug. 4, 1863	39,442–39,445	Destouy, A.	July 31, 1866	56,729
(4 patents on 1 machine)			Schwalback, M.	July 31, 1866	56,805
Tracy and Hobbs	Sept. 15, 1863	40,000	Cately, William H.	Aug. 7, 1866	56,902
Wagener, Jeptha A.	Oct. 13, 1863	40,296	Piper, D. B.	Aug. 7, 1866	56,990
Rehfuss, G.	Oct. 13, 1863	40,311	Leyden, Austin	Aug. 14, 1866	57,157
Lathrop, Lebbeus W., and de Sanno, Wm. P.	Oct. 27, 1863	40,446	Clements, James M.	Aug. 21, 1866	57,451
			Davis, Job A.	Oct. 9, 1866	58,614
Heyer, W. D.	Nov. 17, 1863	40,622	Rodier, Peter	Nov. 13, 1866	59,659
Simmons, A. G., and Scofield, C.	March 1, 1864	11,790	Duchcmin, Wm.	Nov. 13, 1866	59,715
			Kilbourn, E. E.	Nov. 20, 1866	59,746
Guinness, W. S.	March 15, 1864	41,916	Reed, T. K.	Dec. 4, 1866	60,241
Willcox, Charles H.	March 22, 1864	42,036	Singer, I. M.	Dec. 11, 1866	60,433
(4 patents on 1 machine)	Aug. 9, 1864	43,819	Bartram, W. B.	Jan. 1, 1867	60,669
	Sept. 27, 1864	44,490	Rehfuss, G.	Jan. 8, 1867	61,102
	Sept. 27, 1864	44,491	Singer, Isaac	Jan. 15, 1867	61,270
Sibley, J. J.	March 29, 1864	42,117	Cajar, Emil	Feb. 5, 1867	61,711
Thompson, R.	April 19, 1864	42,449	Craige, E. H.	Feb. 19, 1867	62,186
McKay & Blake	May 24, 1864	42,916	Reed, T. K.	Feb. 19, 1867	62,287
Chittenden, H. H.	June 28, 1864	43,289	Bartram, W. B.	March 5, 1867	62,520
Hall, Luther	July 5, 1864	43,404	Fuller, H. W.	March 19, 1867	63,033
Planer, Louis	Aug. 23, 1864	43,927	Stannard, M.	April 23, 1867	64,184
Atwater, B.	Sept. 6, 1864	44,063	Craige, E. H.	Aug. 13, 1867	67,635
Dale, John D.	Oct. 11, 1864	44,686	Doll, Arnold	Sept. 3, 1867	68,420
Gritzner, M. C.	Oct. 18, 1864	44,720	Bruen, L. B.	Sept. 17, 1867	68,839
Smith, DeWitt C.	Dec. 20, 1864	45,528	Hodgkins, C.	Oct. 8, 1867	69,666
Weitling, W.	Jan. 3, 1865	45,777	Baker, G. W.	Oct. 29, 1867	70,152
Cadwell, C.	Jan. 24, 1865	45,972	Cadwell, Caleb	Nov. 19, 1867	71,131
Bartlett, J. W.	Jan. 31, 1865	46,064	Fanning, J.	Dec. 31, 1867	72,829
McCurdy, James S.	Feb. 7, 1865	46,303	Warth, Albin	Jan. 7, 1868	73,064
Lamb, Thomas, and Allen, John	Aug. 15, 1865	49,421	Rehfuss, George	Jan. 7, 1868	73,119
			Cornely, E.	Jan. 28, 1868	73,696
Humphrey, D. W. G.	Aug. 29, 1865	49,627	Blake, L. R.	Feb. 11, 1868	74,289
Tarbox, John N.	Sept. 5, 1865	49,803	Fales, J. F.	Feb. 11, 1868	74,328
Crosby, C. O.	Oct. 3, 1865	50,225	Jencks, G. L.	Feb. 18, 1868	74,694
Cajar, E.	Oct. 3, 1865	50,299	Clark, Edwin E.	Feb. 25, 1868	74,751
Hart, William	Oct. 17, 1865	50,469	Halbert, A. W.	March 31, 1868	76,076
Hecht, A.	Oct. 17, 1865	50,473	Gritzner, M. C.	April 7, 1868	76,323
Emerson, John	Nov. 14, 1865	50,989	Bartlett, Joseph W.	April 7, 1868	76,385
Keats, John, and Clark, Wm. S.	Nov. 14, 1865	50,995	Waterbury, Enos	June 16, 1868	79,037
			Cole, W. H.	June 30, 1868	79,447

Patentee	Date	Patent Number	Patentee	Date	Patent Number
Lamson, Henry P.	July 7, 1868	79,579	Brown, F. H.	April 26, 1870	102,366
French, S.	July 28, 1868	80,345	Howard E., and	May 31, 1870	103,745
Stein, M. J.	Sept. 8, 1868	81,956	Jackson, W. H.		
Hancock, H. J.	Oct. 27, 1868	83,492	Bartram, W. B.	June 14, 1870	104,247
Bartram, W. B.	Nov. 3, 1868	83,592	Henriksen, H. P.	June 21, 1870	104,590
Benedict, C. P.	Nov. 3, 1868	83,596	Martine, Charles F.	June 21, 1870	104,612
Bonnaz, A.	Nov. 10, 1868	83,909	Nasch, Isidor	June 21, 1870	104,630
Bonnaz, A.	Nov. 10, 1868	83,910	Hall, L.	July 12, 1870	105,329
Elliott, F.	Jan. 19, 1869	85,918	Lyon, Lucius	July 26, 1870	105,820
Canfield, F. P.	Jan. 19, 1869	86,057	Bennor, Joseph	Aug. 9, 1870	106,249
Arnold B.	Jan. 26, 1869	86,121	Barnes, M. M.	Aug. 16, 1870	106,307
Jones, John	Jan. 26, 1869	86,163	Leslie, Arthur M.	Oct. 18, 1870	108,492
Russell, W. W.	Feb. 9, 1869	86,695	Rayer, William A., and	Nov. 1, 1870	108,827
Eldridge, G. W.	March 2, 1869	87,331	Lincoln, Wm. S.		
House, J. A. and H. A.	March 2, 1869	87,338	Landfear, Wm. R.	Nov. 22, 1870	109,427
Gird, E. D.	March 9, 1869	87,559	Parham, Charles	Nov. 22, 1870	109,443
Carpenter, William	March 9, 1869	87,633	Lamb, I. W.	Nov. 29, 1870	109,632
Dunbar, C. F.	March 30, 1869	88,282	Moreau, Eugene	Jan. 3, 1871	110,669
McLean, J. N.	March 30, 1869	88,499	Robinson, Charles E.	Jan. 3, 1871	110,790
Billings, C. E.	April 6, 1869	88,603	Goodyear, Charles, Jr.	Jan. 24, 1871	111,197
Winter, Wm.	April 13, 1869	88,936	Stevens, G., and Hendy, J.	Jan. 31, 1871	111,488
Tittman, A.	April 20, 1869	89,093	Carpenter, Mary P.	Feb. 21, 1871	112,016
Swartwout, H. L.	April 27, 1869	89,357	Hancock, Henry J.	Feb. 21, 1871	112,033
Lyons, Lucius	April 27, 1869	89,489	Sidenberg, W.	March 14, 1871	112,745
Crosby, C. O.	May 25, 1869	90,507	Chase, M.	April 11, 1871	113,498
Gutmann, J.	May 25, 1869	90,528	Stein, M. J.	April 11, 1871	113,593
Duchemin, William	June 8, 1869	91,101	Tate, Wm. J.	April 11, 1871	113,704
Adams, John Q.	July 6, 1869	92,138	House, J. A. and H. A.	May 2, 1871	114,294
Bond, Joseph, Jr.	Aug. 10, 1869	93,588	Sidenberg, W.	May 23, 1871	115,117
Hoffman, Geo. W.	Aug. 24, 1869	94,112	Beuttels, Charles	May 23, 1871	115,155
Brown, John H.	Aug. 31, 1869	94,389	Thompson, G.	May 23, 1871	115,255
Heery, Luke	Sept. 14, 1869	94,740	Willcox and Carleton	June 27, 1871	116,521
Gray, Joshua	Oct. 5, 1869	95,581	(3 patents on 1 machine)		116,522
Smith, E. H.	Oct. 26, 1869	96,160			116,523
Page, Chas.	Nov. 2, 1869	96,343	Willcox and Carleton	July 4, 1871	116,783
Lyon, Lucius	Nov. 9, 1869	96,713	Goodyear, Charles, Jr.	July 11, 1871	116,947
Clever, P. J.	Nov. 16, 1869	96,886	Necker, Carl	July 18, 1871	117,101
Mills, Daniel	Nov. 16, 1869	96,944	Pitt, James; Joseph;	July 18, 1871	117,203
Woodruff, Geo. B., and	Nov. 16, 1869	97,014	Edward; and Wm.		
Browning, Geo.			Jones, John T.	Aug. 1, 1871	117,640
Keith, Jeremiah	Dec. 7, 1869	97,518	West, E. P.	Aug. 1, 1871	117,708
Hurtu, Auguste J., and	Dec. 21, 1869	98,064	Jones, Solomon	Aug. 29, 1871	118,537
Hautin, Victor J.			(2 patents on 1 machine)		118,538
Lamb, Thomas	Dec. 28, 1869	98,390	Lamb, Thomas	Sept. 5, 1871	118,728
Rudolph, B.	Feb. 1, 1870	99,481	Bosworth, C. F.	Jan. 9, 1872	122,555
Porter, Alonzo	Feb. 8, 1870	99,704	Smyth, D. M.	Jan. 9, 1872	122,673
Smith, W. T.	Feb. 8, 1870	99,743	Fish, Warren L.	Feb. 13, 1872	123,625
Meyers, N.	Feb. 15, 1870	99,783	Palmer, C. H.	March 19, 1872	124,694
Grover, W. O.	Feb. 22, 1870	100,139	Baker, G. W.	April 9, 1872	125,374
Spoehr, F.	April 12, 1870	101,779	Gordon and Kinert	April 16, 1872	125,807
Kendall, George F.	April 12, 1870	101,887	Howard, C. W.	April 23, 1872	126,056
Cooney, W.	April 26, 1870	102,226	(second machine)		126,057

Patentee	Date	Patent Number
Smyth, D. M.	May 14, 1872	126,845
Beckwith, W. G.	May 21, 1872	126,921
Bouscay, Eloi, Jr.	May 28, 1872	127,145
Braundbeck, E.	June 11, 1872	127,675
Heidenthal, W.	June 11, 1872	127,765
Cleminshaw, S.	June 25, 1872	128,363
Wardwell, S. W., Jr.	July 2, 1872	128,684
Springer, W. A.	July 9, 1872	128,919
Fanning, John	July 16, 1872	129,013
Parks, Volney	July 30, 1872	129,981
Baker, G. W.	July 30, 1872	130,005
Smyth, D. M.	Aug. 6, 1872	130,324
McClure, A. T.	Aug. 13, 1872	130,385
Ashe, Robert	Aug. 20, 1872	130,555
Bartram, W. B.	Aug. 20, 1872	130,557
West, Elliot P.	Aug. 20, 1872	130,674
Happe, J., and Newman, W.	Aug. 20, 1872	130,715
Hinds, Jesse L.	Sept. 10, 1872	131,166
Brown, F. H.	Oct. 1, 1872	131,735
Beckwith, W. G.	Nov. 26, 1872	133,351
Turner, S. S.	Dec. 3, 1872	133,553
Chandler, R.	Dec. 10, 1872	133,757
Venner, O.	Dec. 10, 1872	133,814
Duchemin, W.	Jan. 21, 1873	135,032
Sheffield, G. V.	Jan. 21, 1873	135,047
Parham, Charles	Feb. 4, 1873	135,579
Goodes, E. A.	March 11, 1873	136,718
Tittman, A.	March 11, 1873	136,792
Happe, J., and Newman, W.	March 25, 1873	137,199
Ragan, Daniel	April 1, 1873	137,321
O'Neil, John	April 8, 1873	137,618
Kallmeyer, G.	April 8, 1873	137,689
Ross, J. G., and Miller, T. L.	May 13, 1873	138,764
West, Elliott P.	May 13, 1873	138,772
Koch and Brass	May 13, 1873	138,898
Arnold, B.	May 20, 1873	138,981
Arnold, B.	May 20, 1873	138,982
Lathrop, L. W.	May 20, 1873	139,067
Chandler, Rufus	May 27, 1873	139,368
Jones, S. H.	July 8, 1873	140,631
Smyth, D. M.	July 22, 1873	141,088
Wardwell, S. W., Jr.	July 29, 1873	141,245
Stewart, J., Jr.	July 29, 1873	141,397
Walker, William	July 29, 1873	141,407
Blanchard, Helen A.	Aug. 19, 1873	141,987
Springer, W. A.	Aug. 26, 1873	142,290
Cushman, C. S.	Sept. 2, 1873	142,442
Porter, D. A.	Nov. 25, 1873	144,864
Koch & Brass	Dec. 2, 1873	145,215
Richardson, E. F.	Dec. 16, 1873	145,687
Weber, Theo. A.	Dec. 23, 1873	145,823
Scribner, Benjamin, Jr.	Jan. 13, 1874	146,483
Black, Samuel S.	Jan. 20, 1874	146,642
Taylor, F. B.	Jan. 20, 1874	146,721
Richardson, Everett P.	Jan. 27, 1874	146,948
Muir, William	Feb. 3, 1874	147,152
Goodes, E. A.	Feb. 10, 1874	147,387
Springer, Wm. A.	Feb. 10, 1874	147,441
True, C. B.	March 10, 1874	148,336
Wardwell, S. W., Jr.	March 10, 1874	148,339
Shorey, Samuel W.	March 17, 1874	148,765
Smith, James H.	March 24, 1874	148,902
Horr, Addison D.	April 21, 1874	149,862
Page, Chas.	May 5, 1874	150,479
Crane, Thomas	May 5, 1874	150,532
Buhr, J.	May 26, 1874	151,272
Smyth, D. M.	June 9, 1874	151,801
Wensley, James	June 16, 1874	152,055
Dinsmore, A. S., and Carter, John T.	Jan. 30, 1874	152,618
Speirs, J.	July 7, 1874	152,813
Brewer, A. G.	July 14, 1874	152,894
Baglin, Wm.	Aug. 18, 1874	154,113
Howard, E. L.	Aug. 25, 1874	154,485
Landfear, Wm. R.	Sept. 22, 1874	155,193
Drake, Ellis	Oct. 13, 1874	155,932
Barney, Samuel C.	Oct. 20, 1874	156,119
Moreau, Eugene	Oct. 20, 1874	156,171
Huntington, Thomas S.	Dec. 29, 1874	158,214
Bartlett and Plant	Jan. 26, 1875	159,065
Garland, H. P.	Feb. 16, 1875	159,812
Dinsmore, Alfred S.	March 9, 1875	160,512
McCloskey, John	March 30, 1875	161,534
Schmidt, Albert E.	April 27, 1875	162,697
Darling & Darling	May 25, 1875	163,639
Richardson, Everett P.	July 13, 1875	165,506
Whitehill, Robert	July 27, 1875	166,172
Weber, Theodore A.	Aug. 3, 1875	166,236
Pearson, Wm.	Aug. 17, 1875	166,805
Beckwith, William G.	Sept. 7, 1875	167,382
Hall, John S.	Oct. 11, 1875	168,637
Jones, J. T.	Oct. 26, 1875	169,106
Garland, H. P.	Oct. 26, 1875	169,163
Wormald & Dobson	Nov. 9, 1875	169,881
Rose, R. M.	Nov. 30, 1875	170,596
Keith, Jeremiah	Dec. 7, 1875	170,741
Keith, T. K.	Dec. 14, 1875	170,955
Leavitte, Albert	Dec. 14, 1875	171,147
Toll, Charles F.	Dec. 14, 1875	171,193
Keats, Greenwood, & Keats	Dec. 28, 1875	171,622

Patentee	Date	Patent Number	Patentee	Date	Patent Number
Thayer, Augustus	Jan. 11, 1876	172,205	Beck, August	Nov. 6, 1877	196,863
Frese, B.	Jan. 18, 1876	172,308	Keith, T. H.	Nov. 6, 1877	196,909
Pearson, William	Jan. 18, 1876	172,478	Keats, John	Dec. 11, 1877	198,120
Sawyer & Esty	Feb. 29, 1876	174,159	Briggs, Thomas	Jan. 1, 1878	198,790
Porter & Baker	March 14, 1876	174,703	Corey, J. W.	Jan. 8, 1878	198,970
Walker, William	April 11, 1876	176,101	Howard, T. S. L.	Jan. 15, 1878	199,206
Upson, L. A.	April 18, 1876	176,153	Bosworth, C. F.	Jan. 22, 1878	199,500
Witherspoon, S. A.	April 18, 1876	176,211	Dancel, C.	Jan. 29, 1878	199,802
Rice, T. M.	April 25, 1876	176,686	Pearson, M. H.	Feb. 5, 1878	199,991
Murphy, E.	May 2, 1876	176,880	Morrell, Robert W.; Parkinson, Thomas; and Parkinson, Joseph	April 23, 1878	202,857
Bradford, E. F., and Pierce, V. R.	May 16, 1876	177,371			
Applegate & Webb	May 25, 1876	177,784	Barcellos, D.	April 30, 1878	203,102
Sullivan, John J.	June 27, 1876	179,232	Elderfield, F. D.	June 4, 1878	204,429
Appleton, C. J., and Sibley, J. J.	July 4, 1876	179,440	Heberling, J.	June 4, 1878	204,604
			Beukler, William	June 11, 1878	204,704
Marin, Chas.	July 11, 1876	179,709	Varicas, L.	June 11, 1878	204,864
Gullransen, P. E., and Rettinger, J. C.	July 25, 1876	180,225	Stewart, W. T.	July 2, 1878	205,698
			House, Jas. A.	July 23, 1878	206,239
Butcher, Joseph	Aug. 1, 1876	180,542	Martin, W., Jr.; Dawson, D. R.; and Orchar, R.	Aug. 6, 1878	206,743
Jackson, William	Sept. 5, 1876	181,941			
Barton, Kate C.	Sept. 12, 1876	182,096	Conklin, N. A.	Aug. 6, 1878	206,774
Eickemeyer, Rudolf	Sept. 12, 1876	182,182	Wollenberg, H., and Priesner, J.	Aug. 6, 1878	206,848
Webster, W.	Sept. 12, 1876	182,249			
Knoch, C. F.	Oct. 17, 1876	183,400	Young, E. S., and Dimond, G. H.	Aug. 13, 1878	206,992
Cushman, C. S.	Nov. 21, 1876	184,594			
Harris, David	Dec. 12, 1876	185,228	Hoffman, Clara P., and Meyers, Nicholas	Aug. 13, 1878	207,035
Wood, J.	Dec. 26, 1876	185,811			
Oram, Henry	Jan. 2, 1877	185,952	Wensley, Jas.	Aug. 20, 1878	207,230
Palmer, Frank L.	Jan. 2, 1877	185,954	Dimond, G. H.	Aug. 27, 1878	207,400
Hall, John S.	Feb. 6, 1877	187,006	Steward, A.	Aug. 27, 1878	207,454
Palmateer, William A.	Feb. 20, 1877	187,479	Wood, Richard G.	Sept. 10, 1878	207,928
Cummins, William G.	Feb. 27, 1877	187,822	McCombs, Geo. F.	Sept. 24, 1878	208,407
Esty, William	Feb. 27, 1877	187,837	Keith, Jeremiah	Oct. 22, 1878	209,126
Leavitt & Drew	Feb. 27, 1877	187,874	Wells, W. W.	Nov. 12, 1878	209,843
Henriksen, H. P.	March 20, 1877	188,515	Bayley, C. H.	Feb. 11, 1879	212,122
McKay, Gordon	March 27, 1877	188,809	Parmenter, Charles O.	Feb. 18, 1879	212,495
Follett, J. L.	April 10, 1877	189,446	Ingalls, N., Jr.	Feb. 25, 1879	212,602
Bond, James, Jr.	April 17, 1877	189,599	Cleminshaw, S.	March 18, 1879	213,391
Jacob, F.	April 24, 1877	190,047	Webb, T., and Heartfield, C. H.	March 25, 1879	213,537
Beck, A.	May 1, 1877	190,184			
Hallett, H. H.	June 5, 1877	191,584	Borton, Stockton	April 8, 1879	214,089
Randel, William	June 12, 1877	192,008	Henriksen, H. P.	May 20, 1879	215,615
Corbett, E., and Harlow, C. F.	July 3, 1877	192,568	Bland, Henry	June 3, 1879	216,016
Brown, F. H.	July 24, 1877	193,477	Morrison, T. W.	June 10, 1879	216,289
Melhuish, R. M.	Aug. 28, 1877	194,610	Bosworth, Charles F.	June 17, 1879	216,504
Atwood, K. C.	Sept. 4, 1877	194,759	Simmons, Frederick	June 24, 1879	216,902
Macaulay, F. A.	Oct. 9, 1877	195,939	Junker, Carl	July 1, 1879	217,112
Dimond, George H.	Oct. 16, 1877	196,198	Legat, Désiré Mathurin	Aug. 12, 1879	218,388
Sedmihradsky, A. J.	Oct. 23, 1877	196,486	Willcox, C. H.	Aug. 12, 1879	218,413
Keith, J.	Nov. 6, 1877	196,809	Cornely, Emile	Sept. 2, 1879	219,225

Patentee	Date	Patent Number
Hamm, E.	Sept. 16, 1879	219,578
Tuttle, J. W., and Keith, T. K.	Sept. 16, 1879	219,782
Stackpole, G., and Applegate, J. H.	Oct. 7, 1879	220,314
Otis, S. L.	Oct. 28, 1879	221,093
Bland, H.	Nov. 11, 1879	221,505
Bracher, T. W.	Nov. 11, 1879	221,508
Snediker, J. F.	Nov. 25, 1879	222,089
Mooney, J. H.	Dec. 2, 1879	222,298
Osborne, J. H.	Feb. 3, 1880	224,219
Smith, W. M.	March 2, 1880	225,199
Banks, C. M.	March 23, 1880	225,784
Haberling, J.	May 4, 1880	227,249
Haberling, J.	May 11, 1880	227,525
Wiseman, Edmund	June 8, 1880	228,711
Juengst, George	June 15, 1880	228,820
Morley, J. H.	June 15, 1880	228,918
Curtis, G. H. W.	June 22, 1880	228,985
Lipe, C. E.	June 29, 1880	229,322
Miller, L. B., and Diehl, P.	July 6, 1880	229,629
Willcox, C. H.	July 20, 1880	230,212
Shaw, E.	July 27, 1880	230,580
Dinsmore, A. S.	Aug. 17, 1880	231,155
Thurston, C. H.	Oct. 12, 1880	231,300
Butcher, J.	Oct. 26, 1880	233,657
Smyth, D. M.	Nov. 23, 1880	234,732
Hesse, J.	Dec. 7, 1880	235,085
Kjalman, H. N.	Dec. 21, 1880	235,783
Morley, J. H.	Jan. 4, 1881	236,350
Thomas, J.	Jan. 11, 1881	236,466
Benson, G.	March 8, 1881	238,556
Green, G. F.	March 8, 1881	238,678
Eickemeyer, Rudolf	March 29, 1881	239,319
Palmer, C. H.	April 26, 1881	240,758
Campbell, D. H.	May 17, 1881	241,612
Campbell, Duncan H.	May 17, 1881	241,613
Leslie, A. M.	May 24, 1881	241,808
Newell, George F.	June 7, 1881	242,470
Gritzner, Max C.	June 28, 1881	243,444
Keith, Jeremiah	July 5, 1881	243,710
Choquette, A. E.	July 12, 1881	244,033
Mooney, J. H.	July 19, 1881	244,470
Beardslee, W. F.	Aug. 16, 1881	245,781
Hine, Charlie M.	Aug. 23, 1881	246,136
Willcox, C. H.	Sept. 6, 1881	246,700
Hoefler, J.	Sept. 13, 1881	246,883
Woodward, E.	Sept. 20, 1881	247,285
Richards, Jean E.	Jan. 24, 1882	252,799
Abbott, W. W.	Jan. 31, 1882	252,984
Secor, J. B.	Feb. 14, 1882	253,772
Deschamps, O. L.	Feb. 21, 1882	253,915
Hull, E. H.	Feb. 28, 1882	254,217
Roberts, William	March 7, 1882	254,696
Willcox and Borton	March 28, 1882	255,576
Borton and Willcox	March 28, 1882	255,577
Borton and Willcox	March 28, 1882	255,580
Borton and Willcox	March 28, 1882	255,581
Veukler, W.	April 4, 1882	255,916
Hurtu, A. J.	May 30, 1882	258,761
Keats, Alphonso	July 11, 1882	260,990
Ramsden, John W.	Aug. 1, 1882	262,116
Koch, William	Aug. 8, 1882	262,298
Bigelow, J.	Aug. 29, 1882	263,467
Mills, Daniel	Oct. 10, 1882	265,850
Wilkinson, Charles E.	Dec. 19, 1882	269,251
Carlisle, W. S.	Jan. 9, 1883	270,540
Holden, O. J., and Griswold, L.	Feb. 13, 1883	272,050
Cameron, James W.	Feb. 20, 1883	272,527
Miller, L. B., and Diehl, P.	March 20, 1883	274,359
Ludeke, W.	April 10, 1883	275,506
Bolton, J., and Petnz, A. D.	May 8, 1883	277,106
Blodgett, John W.	June 12, 1883	279,320
Haberling, J.	Sept. 4, 1883	284,300
Thimonnier, E., and Vernaz, C.	Oct. 30, 1883	287,592
Duchemin, William	Nov. 20, 1883	288,929
Lawrence, G. H.	Dec. 25, 1883	290,895
Clever, Peter J.	April 8, 1884	296,529
Palmer, John H.	May 6, 1884	298,228
Dowling, James, and Connolly, John	May 27, 1884	299,118
Boecher, Adam	June 10, 1884	300,199
Luedeke, Waldemar	June 17, 1884	300,380
VanVechten, Orville R.	July 15, 1884	302,063
Carr, Wm. H., and Ostrom, F. W.	Aug. 12, 1884	303,361
Trip, J.	Dec. 2, 1884	308,711
Farrar, Arthur	Dec. 30, 1884	309,837
Turner, M. G.	Feb. 17, 1885	312,306
Mills, D.	March 3, 1885	313,359
Hurtu, August J.	April 7, 1885	315,037
Charmbury, Henry	April 28, 1885	316,745
Woodward & Keith	April 28, 1885	316,927
Walker, William	June 16, 1885	320,099
Tucker, R. D.	June 23, 1885	320,898
Wheeler and Dial	Oct. 13, 1885	328,165
Thomas, Joseph	Nov. 10, 1885	330,170
Muegge, C. A.	Dec. 8, 1885	332,207
Diehl, P.	April 13, 1886	339,623
Diehl, P.	Aug. 24, 1886	347,776

Patentee	Date	Patent Number	Patentee	Date	Patent Number
Helwig, Arthur	Oct. 5, 1886	350,364	Stewart, James, Jr.	July 15, 1890	432,449
Miehling, Charles	Nov. 2, 1886	351,992	Dewees, J. W.	July 22, 1890	432,746
Dieterle, H. E.	Nov. 30, 1886	353,542	Powell, Thomas	Dec. 16, 1890	442,695
Walker, William	Dec. 7, 1886	353,720	Fletcher, James H.	Dec. 30, 1890	443,756
Rosenthal, S. A.	Dec. 7, 1886	353,970	Rudolph, Ernst B., deceased, Boulter, W. E., administrator	April 7, 1891	449,927
Temple, John	Feb. 22, 1887	358,088			
Gee, W. V.	April 19, 1887	361,406			
Lingley, John W.	Aug. 16, 1887	368,538	Goodwin, Julius C.	April 21, 1891	450,793
Boppel, Jacob	Jan. 29, 1889	396,979	Cook, Hugo	June 23, 1891	454,610
Webster, William	April 30, 1889	402,497	Bowyer, J. T.	June 23, 1891	454,708
Osterhout and Hallenbeck	May 7, 1889	402,610	Willcox, C. H., and Borton, S.	April 5, 1892	472,094
Bennett and Dowling	Aug. 27, 1889	409,728			
Hine, Charles M.	Jan. 28, 1890	420,382	Legg and Weston	May 17, 1892	474,840
Wheeler, Nathaniel	Feb. 4, 1890	420,847	Kern, Ferdinand	July 19, 1892	479,369
Hallenbeck, J. P.	April 8, 1890	425,422	Jackson, Francis	May 1, 1894	519,064
Lisle, Myron C.	May 20, 1890	428,171	Abercrombi, Charles	June 5, 1894	520,977
Walker and Bennet	May 20, 1890	428,548	Taft, J. C.	Oct. 15, 1895	547,866

VI. 19th-Century Sewing-Machine Leaflets
In the Smithsonian Collections

Machine or Manufacturer	Date	Type
American B.H.O. and Sewing Machine	1874	Illustrated, advertising leaflet
Buckeye sewing machine	ca. 1870	Illustrated, directions for using the machine
New Buckeye	ca. 1872	Illustrated, directions for using the machine
Centennial sewing machine	1876	Illustrated, advertising leaflet
Domestic sewing machine	1872	Illustrated, advertising leaflet
Florence sewing machine	1873	Illustrated, advertising leaflet
Florence sewing machine	1878	Illustrated, directions for using the machine
Goodes sewing machine	ca. 1876	Advertising leaflet
Grant Brothers sewing machine	1867	Illustrated, advertising leaflet (Xerox copy)
Grover, Baker & Co's Patent Sewing Machines	1853	Illustrated, advertising brochure
Grover and Baker sewing machine	ca. 1870	Illustrated, advertising leaflet
Home sewing machine	ca. 1870	Illustrated, advertising leaflet
Howe sewing machine, new "B" machine	1868	Illustrated, instruction booklet
Howe sewing machine	1876	Illustrated, catalog of machines
Independent Noiseless sewing machine	ca. 1874	Illustrated, advertising leaflet
Ladd, Webster sewing machine	1861	Illustrated, advertising leaflet
Little Monitor sewing machine	ca. 1872	Illustrated, advertising leaflet
Remington Family sewing machine	ca. 1874	Illustrated, advertising leaflet
Shaw and Clark sewing machine	1864	Illustrated, advertising leaflet
Singer sewing machine	1871	Illustrated, advertising leaflet
Singer sewing machine	1893	Catalog of machines shown at the Columbian Exposition
Singer Vibrating Shuttle Machine	1889	Directions for use
Standard Shuttle sewing machine	ca. 1875	Illustrated, advertising leaflet
Ten Dollar Novelty sewing machine	ca. 1870	Illustrated, advertising leaflet
Weed sewing machine	1873	Illustrated, advertising leaflet
Wheeler & Wilson Sewing Machine Co.	1863	Advertising brochure with history of sewing machine
Wheeler and Wilson sewing machine	ca. 1869	Illustrated, instruction booklet
Wheeler and Wilson sewing machine	ca. 1870–1875	Illustrated, advertising leaflet
Wheeler and Wilson no. 8 machine	ca. 1878	Illustrated, instruction booklet
Willcox & Gibbs Noiseless Family Sewing Machine	ca. 1864	Illustrated advertising booklet
Willcox & Gibbs Noiseless Family Sewing Machine	ca. 1868	Illustrated advertising booklet
Wilson sewing machine	1872	Illustrated, advertising leaflet

VII. A Brief History of Cotton Thread

Although Samuel Slater's wife is credited with making the first cotton sewing thread from yarns spun at the Pawtucket, Rhode Island, mill in about 1794, cotton thread did not become a manufactured item at that time. Slater turned all his interests to producing cotton-twist yarns needed for the warps of cotton fabrics. By 1809, however, the agents of Almy and Brown, partners and distributors for Slater, were advertising cotton thread as follows:

> Factory Cotton and Thread Store, No. 26 Court Street opposite Concert Hall. George Connell, Agent for Almy and Brown of Providence and Pawtucket Manufactories, has now for sale from eight to ten thousand weight of yarn, for weaving . . . five hundred pounds cotton thread, in hanks, from No. 12 to 60 of a superior quality and very white.*

Although it was a short hop from the spinning of cotton warps to the twisting of these cotton yarns to form a sewing thread, the general manufacture of cotton thread as an industry did not originate in the United States but rather in Scotland in the early 19th century. Napoleon's blockade, which curtailed Great Britain's importation of silk—needed not only for fabrics but also for making heddle strings for the looms—stimulated the production of cotton thread there. James and Patrick Clark, in desperation, attempted to substitute cotton for silk in their manufacture of these heddle strings. When they were successful, they considered that if cotton could be used successfully for this purpose it could also be made suitable for sewing thread. In 1812 they built a factory in Paisley, Scotland, which had long been noted for its textile industries. The thread was sold in hanks. About 1820 James' sons, James and John, who were now running J. & J. Clark & Co., began to wind the thread on spools. For this service they charged an extra halfpenny, which was refunded when the empty spool was returned. The thread was usually a three-ply or so-called three-cord thread.

About 1815 James Coats, also of Paisley, started manufacturing thread at Ferguslie, Scotland. His two sons took over the company in 1826 and formed the J. & P. Coats Company. Another brother, Andrew Coats, became the selling agent in the United States about 1840. But the cotton-thread industry was not fully launched.

As reported in an 1853 *Scientific American*, there was "more American thread made ten years ago than there is today."** It was not until the six-cord cabled cotton thread, which was suitable for both machine and hand sewing, was perfected that the industry progressed into full operation.

---

*WILLIAM R. BAGNALL, *Textile Industries of the United States* (Cambridge, Mass., 1893), vol. 1, p. 164.

**Scientific American* (Oct. 22, 1853), vol. 9, no. 6, p. 46.

VIII. Biographical Sketches

BARTHELEMY THIMONNIER

The first man known to have put a sewing machine into practical operation, Barthelemy Thimonnier, was a Frenchman of obscure parentage. His father, a textile dyer of Lyon, left that city in 1793 as a result of the Revolution and journeyed with his family to l'Arbresle where Barthelemy was born in August of that year.

The family resources were small, and, although the young Thimonnier was able to begin studies at the Seminaire de Saint-Jean at Lyons, he soon was forced to leave school for financial reasons and return to his home, then at Amplepuis. There he learned the tailoring trade and by 1813 was fairly well established in his own shop.

At that time many of the town's inhabitants were weavers and almost every house possessed one or two looms. The noise of the shuttle echoed from these family workshops. Thimonnier noted the relatively small amount of time needed to weave a fabric compared with the slow painstaking task of sewing a garment by passing the needle in and out for each stitch of each seam. When his mind began to dwell on the idea of producing a machine to do this stitching, another of the town's occupations supplied him with a clue and an additional incentive. This village industry produced a type of embroidery work called *point de chainette*, in which a needle with a small hook was used to form the chainstitch, a popular type of decorative stitch long used in countries all over the world. It was Thimonnier's plan to use this type of hooked needle and produce the stitch by machine, employing it both as a decorative stitch and a seam-forming one.

In 1825 Thimonnier moved to St. Etienne, where he became completely absorbed in the idea of inventing a sewing machine. Ignorant of any of the principles of mechanics, he worked alone and in secret for four years, neglecting his tailoring business to the extent that neighbors looked upon him as peculiar, if not crazy. By 1829 he had not only mastered the mechanical difficulties of bringing his dream to realization, but also had made the acquaintance of the man who helped him to success. Ferrand, of l'Ecole des Mines of Saint-Etienne, became interested in the machine and helped finance Thimonnier through his trials and disappointments. In 1830 Thimonnier received a patent on his machine, which produced the chainstitch by means of a needle shaped like a small crochet hook.

Thimonnier, together with Ferrand and a M. Beaunier, made attempts to introduce his machine in Paris. By 1841 they were successful in having eighty of Thimonnier's machines in use sewing army clothing in a shop in Paris. But the fears of the tailors could not be quieted. The machines were destroyed by an ignorant and

Figure 215.—Barthelemy Thimonnier, 1793–1857. From an engraving in the *Sewing Machine Advance*, November 15, 1880. (Smithsonian photo 10569–A.)

infuriated mob, as had been earlier labor-saving devices such as the Jacquard attachment for the loom and Hargreaves' spinning jenny. Thimonnier was forced to flee to his home in St. Etienne, once more penniless.

Soon after this, Jean Marie Magnin, an engineer from Ville-franche-sur-Saône became interested in Thimonnier's machine and provided the inventor again with financial backing. In 1845 under the name of Thimonnier and Magnin the patent of 1830 was renewed, and under it they organized the first French sewing-machine company. The machines they manufactured could produce 200 stitches per minute.

The Revolution of 1848 curtailed the manufacture and sale of the machines. Thimonnier, remembering his unpleasant experience in 1841, decided to go to England with Magnin, where, on February 8, 1848, they received the English patent for his chainstitch machine. He was also granted United States patent 7,622 on September 20, 1850. This later machine had some advantages over his French machine of 1830, but by this time other inventors had joined the field with machines that were more practical. Magnin entered a sewing machine (which from the description in the catalog must have been Thimonnier's invention) in the Crystal Palace Exhibition in London in 1850, but because it was late in arriving it was overlooked by the judges and not even considered in the competition. Thimonnier died in poverty at Amplepuis on July 5, 1857.

WALTER HUNT

Walter Hunt was born near Martinsburg, New York, on July 29, 1796. Although little is known of Hunt's early childhood, we do learn from the author of his obituary, which appeared in *Scientific American*, July 9, 1860, that even as a child he was more interested in people and what he could do for them than in what he could do to insure his own welfare. He is said to have devoted his life to his friends, frequently giving away his last cent when he did not have enough to provide for himself.

There is no record that Hunt maintained a regular business other than the occupation of inventor. His interests were numerous and varied. He received his first patent on June 26, 1826, for a machine for spinning flax and hemp. During the next 33 years he patented 26 ideas. In addition he sold or dropped several more. His second patent was for a coach alarm, and through the years he also received patents for a variety of things including a knife sharpener, heating stove, ice boat, nail machine, inkwell, fountain pen, safety pin, bottle stopper, sewing machine (1854), paper collars, and a reversible metallic heel.

Figure 216.—WALTER HUNT, 1796–1860. From a daguerreotype owned by his great-grandson, C. N. Hunt. (Smithsonian photo 32066–A.)

ELIAS HOWE, JR.

Elias Howe, Jr., was born on his father's farm in Spencer, Massachusetts, on July 9, 1819. This was one of those barren New England farms with many rock-filled acres. All possible ingenuity was necessary to secure a living. The elder Howe supplemented his farming by having a small gristmill, a sawmill, and also by manufacturing cards for the fast-growing cotton industry of New England. Elias Jr.'s earliest recollections were of the latter. He worked with his brothers and sisters sticking wire teeth into strips of leather to make these cotton cards, but, not being very good at this, his family decided to let him "live out" with a neighboring farmer. (Children were leased in those days; they received their board and keep in exchange for chores they would perform.) After a few years, Elias returned home and worked in his father's mill until he was sixteen. Then, against the wishes of his family, he went to Lowell, Massachusetts. Here, he obtained a learner's place in a machine shop where cotton-spinning machinery was made and repaired.

In 1837, when a financial panic hit the country, Howe lost his job. He then decided to go to Boston, and this marked a turning point in his career. In Boston he met Ari Davis, a maker of mariners' instruments and scientific apparatus. Howe began to work in Davis' shop, a place to which inventors often came to ask advice about their ideas. Davis sometimes helped them, but just as often he shouted at them in anger—he is said to have been one of the noisiest men in Boston. One day Howe overheard his employer bellowing at a man who had brought a knitting machine to the shop to seek Davis' advice. "Why are you wasting your time over a knitting machine?" said Davis, "Take my advice, try something that will pay. Make a sewing machine." "It can't be done," was the reply. "Can't be done?" shouted Davis, "Don't tell me that. Why—I can make a sewing machine myself." "If you do," interrupted the capitalist, "I can make an independent fortune for you." Davis, like most men of many words, often talked of more than he planned to do. He never attempted to invent a sewing machine.

But the loud voices interested Howe, who, it is said, determined then that he would produce a sewing machine and win the fortune that the prosperous-looking man had asserted was waiting for such a deed. A kind of lameness since birth had made physical tasks painful for Howe, and he perhaps felt that this would offer an opportunity to become independent of hard physical work.

After marrying on a journeyman machinist's pay of $9 a week, Howe's health worsened and by 1843 was so bad that he had to stop work for days at a time. His wife was forced to take in sewing to maintain the family. It was the sight of his wife toiling at her stitches together with the pressure of poverty that recalled to Howe his earlier interest in a machine to sew. He decided to make an earnest attempt to invent one. Watching his wife for hours at a time, he tried to visualize a machine that would duplicate the motions of the arm. After many trials, he conceived the idea of using an eye-pointed needle in combination with a shuttle to form a stitch. It is possible that, as some authors state, the solution appeared to him in a dream, a manifestation of the subconscious at work. Others have suggested that he may have learned of Hunt's machine. There is a general similarity in the two, not only in the combination of eye-pointed needle and shuttle but in the overhanging arm and vertical cloth suspension.

After conceiving the idea, whatever his inspiration, Howe determined to devote all of his time to producing a working model of his machine. Elias' father, who had then started a factory for splitting palm leaves in Cambridge, gave him permission to set up a lathe and

Figure 217.—ELIAS HOWE, JR., 1819–1867. From an oil painting in the Smithsonian Institution presented by the inventor's grandson, Elias Howe Stockwell. (Smithsonian photo 622.)

a few tools in the garret of the factory. Elias moved his family to Cambridge. Soon after his arrival, unfortunately, the building burned down, and Howe despaired of finding a place to work. He had a friend, however, in George Fisher, who had just come into a small inheritance, and Howe persuaded him to enter into partnership with him for the development of the machine. Fisher agreed to board Howe and his family, which now included two children, while Howe completed the model. Fisher also agreed to supply $500 for material and tools in exchange for a half interest in a patent if one was obtained.

At long last Howe was able to spend his full time and concentration on building his machine. His family was being fed and had a roof over its head. Within a few months Howe had completed a model and by May 1845 had sewed his first seam (see fig. 14). In July of that year he sewed all the principal seams of two suits of wool clothes, one for George Fisher and one for himself.

Several efforts were made to solicit public interest in the new machine. One was installed in a public hall in Boston, and a tailor was employed to operate it at three times the regular wage. The reception was similar to that of Thimonnier's: crowds came to see the "contraption," but, when Howe tried to interest large clothing establishments in using the machine, the protests of the tailors effectively blocked him. Of the several stories about Howe's demonstration of the machine and his use of it to stitch clothing, the one most repeated was first published by James Parton in 1867. In this, Howe had a race with five seamstresses in which he stitched five seams faster than each of them completed one. Parton also refers to Howe as sewing "by his machine all the seams of two suits of woolen clothes." Since Howe's machine could only stitch straight, short seams without resetting the basterplate, the latter story could not be true. Parton probably misinterpreted information given by Howe in his 1860 patent extension appeal. Howe reported he had "tested its practical success by sewing with it all the principal seams in two suits of clothes." Whether Parton's stories were true or not, Howe did not receive a single order. It was estimated a large shirtmaker would have to buy thirty or forty such machines and the investment needed was too large.

Howe was not too discouraged. In the meantime, he had finished a second machine for deposit with the patent specifications, as the patent laws then required. The second was a better made machine (fig. 15) and showed several minor changes. As soon as the patent was issued on September 10, 1846, Howe and his partner returned to Cambridge.

Without the inventor's enthusiasm or love of his own invention, George Fisher became thoroughly discouraged. He had boarded Howe and his family for nearly two years, had furnished the money needed to purchase the tools and materials for making the two sewing machines, had met the expense of obtaining the patent and the trip of Howe and himself to Washington; representing in all an outlay of practically $2000. Since no orders for machines had been received from either garment makers or tailors, Fisher did not see the slightest probability of the machine's becoming profitable and regarded his advances of cash as a dead loss.

Howe moved back to his father's house with a plan to look elsewhere for a chance to introduce the machine. Obtaining a loan from his father, he built another machine and sent it to England by his brother Amasa. After many discouraging attempts to interest the British, Amasa met William Thomas, a manufacturer of umbrellas, corsets, and leather goods. Thomas employed many workmen, all of whom stitched by hand, and he immediately saw the possibilities of a sewing machine. He proposed that Howe sell the machine to him for £250 sterling (about $1250). Thomas further proposed to engage the inventor to adapt this machine to the making of corsets, at a salary of £3 a week.

When Amasa Howe returned to Cambridge with the news, Elias was reluctant to accept Thomas' offer but had nothing better in sight. So the brothers sailed for London in February 1847, taking with them Howe's first machine and his patent papers. Thomas later advanced the passage money for Howe's wife and three children so that they could join Howe in England.

At this point, historians disagree on how long Howe was in Thomas' employ and whether he succeeded in adapting the machine to meet Thomas' needs. He was in England long enough, however, to find himself without employment in a strange country, his funds nearly exhausted, and his wife ill. He hoped to profit by the notice that his work had received and began to build another machine. He sent his family home to reduce expenses while he stayed on to finish the machine.

After working on it for three or four months, he was forced to sell it for five pounds and to take a note for that. To collect enough for his passage home, he sold the note for four pounds cash and pawned his precious first machine and his patent papers. He landed in New York in April 1849 with but half a crown in his pocket to show for his labors. A short time after he arrived, he learned that his wife was desperately ill. Only with a loan from his father was he able to reach her side before she died. Friends were found to look after the children, and Elias returned to work as a journeyman machinist.

Howe discovered, much to his surprise, that during his absence in England the sewing machine had become recognized in the United States. Several machines made in Boston had been sold to manufacturers and were in daily operation. Upon investigating them, he felt that they utilized all or part of the invention that he had patented in 1846, and he prepared to secure just compensation for its use. The first thing he did was to regain his first machine and patent papers from the London pawnshop. It was no easy matter for Howe to raise the money, but by summer he had managed. It was sent to London with Anson Burlingame, who redeemed the loans, and by autumn of the same year the precious possessions were back in Howe's hands. Though Howe gained nothing by his English experience, William Thomas by his modest expenditure obtained all rights to the machine for Great Britain. This later proved to be a valuable property.

Howe then began writing letters to those whom he considered patent infringers, requesting them to pay a fee or discontinue the manufacture of sewing machines which incorporated his patented inventions. Some at first were willing to pay the fee, but they were persuaded by the others to stand with them and resist Howe. This action forced Howe to the courts. With his father's aid he began a suit, but soon found that considerably more money than either possessed was necessary for such actions. Howe turned once more to George Fisher, but years of investing money in Howe's machine without any monetary return had cooled him to the idea. Fisher, however, agreed to sell his half interest, and in February 1851 George S. Jackson, Daniel C. Johnson, and William E. Whiting became joint owners with Howe. These men helped Howe to procure witnesses in the furtherance of numerous suits, but more money was needed than they could raise. The following year a Massachusetts man by the name of George W. Bliss was persuaded to advance the money for the heavy legal expenses needed to protect the patent. Bliss did this as a speculation and demanded additional security. Once more Elias' long-suffering parent came to the rescue and mortgaged his farm to get the necessary collateral.

Only one of these suits was prosecuted to a hearing, but this one, relatively unimportant in itself, set the precedent. In it the defense relied on the earlier invention of Walter Hunt to oppose Howe's claims. The defendant succeeded in proving that Hunt invented, perfected, and sold two machines in 1834 and 1835 which contained all the essential devices in Howe's machine of 1846. But Howe showed that the defendant's machine (which was a Blodgett and Lerow) contained some features of Howe's machine which were not in Hunt's. The jury decided the case in favor of Howe. Howe later fought a vigorous battle with Isaac Singer, but after much legal controversy the ultimate decision in that case also was in Howe's favor. The suits and payments to each patent holder for the right to use his idea were choking the sewing-machine industry. Even Howe could not manufacture a practical machine without an infringement. Finally an agreement was reached and a "Combination" was formed by the major patent holders (see pp. 41–42).

In the meantime, eight years of the first term of Howe's patent had expired without producing much revenue. This permitted Howe, upon the death of his partner, George Bliss, to buy Bliss' half interest for a small sum. He became, then, the sole owner of his patent just as it was to bring him a fortune. He obtained a seven-year extension for his patent in 1860 without any difficulty, and in 1867, when he applied for another extension, he stated that he had received $1,185,000 from it. Though he endeavored to show that because of the machine's great value to the public he was entitled to receive at least $150,000,000, the second application was denied.

During the Civil War, Howe enlisted as a private soldier in the 17th Regiment Connecticut Volunteers. He went into the field and served as an enlisted man. On occasion when the Government was pressed for funds to pay its soldiers, he advanced the money necessary to pay his entire regiment.

Howe did not establish a sewing-machine factory until just before his death in 1867. One of his early licensees had been his elder brother, Amasa, who had organized the Howe Sewing Machine Company about 1853. When Elias began manufacturing machines on his own, he sunk into the bedplate of each machine a brass medallion bearing his likeness. Elias gave his company the same name that his elder brother had used. As this had been Amasa's exclusive property for many years, he took the matter to the courts where the decision went against Elias. He then organized the Howe Machine Company and began to manufacture sewing machines. On October 3, 1867, Elias died in Brooklyn, New York, at the home of one of his sons-in-law. The company was then carried on by his two sons-in-law, who were Stockwell brothers. In 1872 the Howe Sewing Machine Company was sold by Amasa's son to the Stockwells' Howe Machine Company, which in turn went out of business in the mid-1880s.

ALLEN BENJAMIN WILSON

Allen B. Wilson was born in the small town of Willett, Cortlandt County, New York, in 1824. At sixteen he was apprenticed to a distant relative, a cabinetmaker. Unfortunate circumstances caused him to leave this employ, and in 1847 Wilson was in Adrian, Michigan, working as a journeyman cabinetmaker. The place and year are important, for it was at this time that he conceived his idea of a sewing machine. Because of the distant location, it is believed that he was not aware of similar efforts being made in New England. Wilson became ill and for many months could not work at his trade. By August 1848 he was able to work again and found employment at Pittsfield, Massachusetts. Resolving to develop his idea of a sewing machine, he worked diligently and by November had made full drawings of all the parts, according to his previous conceptions.

In comparison to the monetary returns received by the inventors Howe and Singer, Wilson himself did not receive as great a monetary reward for his outstanding sewing-machine inventions. Because of his health Wilson retired in 1853, when the stock company was formed, but he received a regular salary and additional money from the patent renewals. Wilson petitioned for a second extension of his patents on April 7, 1874, stating that, due to his early poverty, he had been compelled to sell a half interest in a patent (his first one) for the sum of $200. Also he stated that he had not received more than his expenses during the original fourteen-year term. Wilson also stated that he had received only $137,000 during the first seven-year extension period. These figures were verified by his partner. The petition was read before both Houses of Congress and referred to the

Figure 218.—ALLEN BENJAMIN WILSON, 1824–1888. From a drawing owned by the Singer Mfg. Co. Formerly, the drawing was owned by the Wheeler & Wilson Mfg. Co. (Smithsonian photo 32066.)

Committee on Patents.* There was strong feeling against the extension of the Wilson patents. The New York *Daily Graphic*, December 30, 1874, reported:

> So valuable has been this latter four-motion feed that few or no cloth-sewing machines are now made without it. The joint ownership of this feature of the Wilson patents has served to bind the combination of sewing-machine builders together, and enabled them to defy competition by force of the monopoly. It is this feature which the combination wishes to further monopolize for seven years by act of Congress. The inventor has probably realized millions for his invention. Singer admits that his patents, which are much less important, paid him two millions prior to 1870, since which time he has not been compelled to render an account. The Wilson patents with their extended terms were worth a much larger sum. They have been public property, so far as the feed is concerned, since June 15, 1873, and will remain so if too great a pressure is not brought to bear on Congress for their extension. A monopoly of this feed motion for seven years more would be worth from ten to thirty millions to the owner—and would cost the people four times as much.

Wilson had not made the millions for he only received a small percentage of the renewals' earnings plus his salary from the patents' owner, the Wheeler and Wilson Manufacturing Company.

The Congressional Committee on Patents made an adverse report in 1874 and again in 1875 and 1876, when applications for an extension were continued.

Wilson died on April 29, 1888.

ISAAC MERRITT SINGER

Isaac Singer, whose name is known around the world as a manufacturer of sewing machines, was the eighth child of poor German immigrants. Isaac was born on October 27, 1811, in Pittstown, New York, but most of

Figure 219.—ISAAC MERRITT SINGER, 1811–1875. From a charcoal drawing owned by the Singer Mfg. Co. (Smithsonian photo 32066–B.)

The Proceedings and Debates of the 43rd Congress, First Session, 1874 Congressional Record, vol. 2, part 3, petition read to the House by Mr. Creamer on April 7, 1874. In part 4 of the same, Mr. Buckingham read a similar petition to the Senate on May 19, 1874. Both were referred to the Committee on Patents; an extension was not granted.

his early life was spent in Oswego. He worked as a mechanic and cabinetmaker, but acquired an interest in the theater. Under the name of Isaac Merritt, he went to Rochester and became an actor. In 1839, during an absence from the theater, he completed his first invention, a mechanical excavator, which he sold for $2000. With the money Singer organized a theatrical troupe of his own, which he called "The Merritt Players." When the group failed in Fredericksburg, Ohio, Singer was stranded for lack of funds.

Forced to find some type of employment, Singer took a job in a Fredericksburg plant that manufactured wooden printers' type. He quickly recognized the need for an improved type-carving machine. After inventing and patenting one, he found no financial support in Fredericksburg and decided to take the machine to New York City. Here, the firm of A. B. Taylor and Co. agreed to furnish the money and give Singer room in its Hague Street factory to build machines. A boiler explosion destroyed the first machine, and Taylor refused to advance more money.

While Singer was with Taylor, George B. Zieber, a bookseller who had seen the type-carving machine, considered its value to publishers. Zieber offered to help Singer and raised $1700 to build another model. In June 1850 the machine was completed. Singer and Zieber took the machine to Boston where they rented display space in the steam-powered workshop of Orson C. Phelps at 19 Harvard Place. Only a few publishers came to look at the machine, and none wanted to buy it.

Singer, contemplating his future, became interested in Phelps' work, manufacturing sewing machines for J. A. Lerow and S. C. Blodgett. Phelps welcomed Singer's interest as the design of the mechanism was faulty and purchasers kept returning the machines for repairs. Singer examined the sewing machine with the eyes of a practical machinist. He criticized the action of the shuttle, which passed around a circle, and the needle bar, which pushed a curved needle horizontally. Singer suggested that the shuttle move to and fro in a straight path and that a straight needle be used vertically. Phelps encouraged Singer to abandon the type-carving machine and turn his energies toward the improvement of the sewing machine. Convinced that he could make his ideas work, Singer sketched a rough draft of his proposed machine, and with the support of Zieber and Phelps the work began.

Singer continued to be active in the sewing-machine business until 1863. He made his home in Paris for a short time and then moved to England. While living at Torquay he conceived the idea of a fabulous Greco-Roman mansion, which he planned to have built at Paignton. Singer called it "The Wigwam." Unfortunately, after all his plans, he did not live to see its completion. Singer died on July 23, 1875, of heart disease at the age of sixty-three.

Bibliography

ADAMS, CHARLES K. Sewing machines. In vol. 7 of *Johnson's universal cyclopaedia*, New York: D. Appleton and Company, 1895.

ALEXANDER, EDWIN P. Sewing, plaiting, and felting machines. Pp. 341–353 in *The Practical Mechanic's Journal's record of the great exhibition, 1862.* 1862.

———. On the sewing machine: Its history and progress. *Journal of the Royal Society of Arts* (April 10, 1863), vol. 2, no. 542, p. 358.

AMERICAN HISTORICAL SOCIETY. *The life and works of George H. Corliss.* (Privately printed for Mary Corliss by the Society.) 1930.

Biography of Isaac M. Singer. *The Atlas*, New York, N.Y., March 20, 1853.

BISHOP, J. LEANDER. In vol. 2 of *A history of American manufactures from 1608 to 1860*, by Bishop; Philadelphia: Edward Young and Company, 1866.

BOLTON, JAMES. Sewing machines, special report on special subjects. In vol. 2 of *Report of the Committee on Awards;* Chicago: World's Columbian Exposition, 1893.

BROCKETT, LINUS P. Sewing machines. In vol. 4 of *Johnson's universal cyclopaedia;* New York: D. Appleton and Company, 1874.

BURR, J. B. *The great industries of the United States.* Hartford: J. B. Burr and Hyde, n.d.

COOK, ROSAMOND C. *Sewing machines.* Peoria, Illinois: Manual Arts Press, 1922.

DEPEW, C. M. 100 years of American commerce. In vol. 2, *American sewing machines* by F. G. Bourne. New York: D. O. Haines, 1895.

DISTRICT OF COLUMBIA CIRCUIT COURT. Case of Hunt vs. Howe. In *Federal Cases*, case 6891, book 12, 1855.

DOOLITTLE, WILLIAM H. *Inventions.* Vol. in Library of modern progress. Philadelphia: Modern Progress Publishing Company, 1905.

DURGIN, CHARLES A. *Digest of patents on sewing machines, Feb. 21, 1842—July 1, 1859.* New York: Livesey Brothers, 1859.

FAIRFIELD, GEORGE A. Report on sewing machines. In vol. 3 of *Report of the commissioners of the United States to the international exhibition held at Vienna, 1873.* 1874.

FISKE, BRADLEY A. *Invention, the master-key to progress.* New York: E. P. Dutton and Company, 1921.

GIFFORD, GEORGE W. [Counsel for Elias Howe, Jr.] *Argument of Gifford in favor of Howe's application for extension of patent.* Washington: U.S. Patent Office, 1860.

GRANGER, JEAN. Thimonnier et la machine à coudre. In vol. 2 of *Les publications techniques.* Paris, 1943. [Source for biographical sketch on Barthelemy Thimonnier.]

GREGORY, GEORGE W. *Machines, etc., used in sewing and making clothing.* Vol. 7 of *Reports and Awards.* Philadelphia International Exhibition, 1876. U.S. Centennial Commission, Washington, D.C., 1880.

HERSBERG, RUDOLPH. *The sewing machine, its history, construction and application.* Transl. Upfield Green. London: E. &. F. N. Spon, 1864.

HOWE, HENRY. *Adventures and achievements of Americans.* Published by the author, 1860.

HOWE MACHINE COMPANY. *The Howe exhibition catalog of sewing machines.* Published by the company for the Philadelphia Centennial Exposition, 1876.

HUBERT, PHILIP G., JR. *Men of achievement: inventors.* New York: Charles Scribner's Sons, 1894.

HUNT, CLINTON N. *Walter Hunt, American inventor.* Published by the author, 1935. [Source for biographical sketch on Walter Hunt.]

INTERNATIONAL EXHIBITION. Sewing and knitting machines. In *Reports of the Juries*, Class VII, Section A. London, 1862.

JOHNSON CLARK AND COMPANY. *Sewing machines for domestic and export trade.* Published by the company, 1883.

KAEMPFFERT, WALDEMAR. A popular history of American invention. *Making clothes by machine*, by John Walker Harrington, vol 2. New York: Charles Scribner's Sons, 1924.

KILGARE, C. B. *Sewing-machines.* Philadelphia: J. B. Lippincott Company, 1892.

KNIGHT, EDWARD H. Vol. 3 of *Knight's American mechanical dictionary.* Boston: Houghton, Mifflin and Company, 1882.

LEWTON, FREDERICK L. The servant in the house: a brief history of the sewing machine. Pp. 559–583 in Annual Report of the Smithsonian Institution (1929); Washington: 1930.

LUTH, ERICH. *Ein Mayener Strumpfwirker, Balthasar Krems, 1760-1813, Erfinder der Nähmaschine.* Hamburg, Germany: Verlag Herbert Luth, 1941.

———. *Josef Madersperger oder der unscheinbore Genius.* Hamburg, Germany: Reichsnerband des Deutschen Nähmaschinenhandels EV, 1933 [?].

Obituary of Isaac M. Singer. *New York Tribune*, July 26, 1875.

MADERSPERGER, JOSEPH. *Beschreibung einer Nähmaschine.* Vienna, ca. 1816.

O'BRIEN, WALTER. Sewing machines. *Textile American*, reprinted March 1931. [A paper read at a meeting of the Leicester Textile Society.]

ORCUTT, REV. SAMUEL. *A history of the old town of Stratford and the city of Bridgeport, Connecticut.* 2 vols. Fairfield County Historical Society, 1886.

PARIS UNIVERSAL EXPOSITION. *Reports of the U.S. Commissioners General Survey*, vol. 1, pp. 293–297. 1867.

PARTON, JAMES. History of the sewing machine. *Atlantic Monthly*, May 1867.

RAYMOND, WILLIAM CHANDLER. *Curiosities of the Patent Office.* Syracuse, N.Y.: published by the author, 1888.

REID, J. A. *A monograph prepared for American industrial education conference and exposition.* Elizabethport, N.J.: Singer Manufacturing Company, 1915.

RENTERS, WILHELM. *Praktisches wissin von der Nähmaschine.* Langensalza, Berlin, and Leipzig: Verlag Julius Beltz, 1933.

Scientific American. Issues of January 27, February 3, and November 24, 1849; July 17, 1852; July 9 and November 5, 1859; January 3 and May 16, 1863; October 19, 1867; and July 25, 1896.

SCOTT, JOHN. *Genius rewarded, or the story of the sewing machine.* New York: J. J. Caulon, 1880.

Sewing Machine Advance. Chicago. Vols. 1–35, no. 5, May 1879 to 1913.

Sewing Machine Journal. September 1879 to December 1884.

Sewing Machine News. New York. Vols. 1–6, 1877–1879; n.s. vols. 1–16, 1879–1893.

Sewing Machine Times. New York. Vols. 1-9, 1882-1890; n.s. vols. 1-38 (nos. 1-725), 1891-April 1924.

STAMBAUGH, JOHN P. *Sewing machines at the 22nd annual exhibition of the Maryland Institute, also a history of the sewing machine.* Hartford, Conn.: Weed Sewing Machine Company, 1872.

Story of sewing machine patents. *New York Tribune,* May 23, 1862.

THOMPSON, HOLLAND. The age of invention. Vol. 37 of the Chronicles of America. New Haven: Yale University Press, 1921.

TOWERS, HENRY M. *Historical sketches relating to Spencer, Massachusetts.* Vol. 1. Spencer, Mass.: W. G. Heffernan, 1901.

URQUHART, J. W. *Sewing machinery.* Vol. 8 in Weale's rudimentary scientific and educational series. London: Crosby Lockwood and Co., 1881.

VAN SLYCK, J. D. *New England manufacturers and manufactories.* Vol. 2. Boston: Van Slyck and Company, 1879. [Source for biographical sketch on Allen Benjamin Wilson, Van Slyck, vol. 2, pp. 672-682.]

WALSO, GEORGE C., JR. *History of Bridgeport and vicinity.* Vol. 2. Bridgeport, Conn.: S. J. Clarke Publishing Company, 1917.

Who invented the sewing machine? *The Galaxy* [article unsigned], vol. 4, August 1867.

Indexes

Geographical Index to Companies Listed in Appendix II

CONNECTICUT

Birmingham
    Wilson H. Smith, 76
Bridgeport
    Avery Mfg. Co., 66
    D.W. Clark, 67
    Jerome B. Secor, 69
    Goodbody Sewing Machine Co., 70
    Howe Machine Co., 70
    New Howe Mfg. Co., 73
    Secor Machine Co., 74
    American Hand Sewing Machine Co., 74
    Singer Co., 76
    Wheeler & Wilson Mfg. Co., 75, 76
Bristol
    Nettleton & Raymond, 74
    Watson & Wooster, 75
Danbury
    Bartram & Fanton Mfg. Co., 66
Hartford
    Morrison, Wilkinson & Co., 72
    T.E. Weed & Co., 75
    Weed Sewing Machine Co., 75
Meriden
    Fosket and Savage, 69
    Charles Parker Co., 73
    Parker Sewing Machine Co., 73
New Britain
    Thurston Mfg. Co., 67
Middletown
    Victor Sewing Machine Co., 75
Norwich
    Greenman & True Mfg. Co., 70
    Morse & True, 70
Wallingford
    Wilson Sewing Machine Co., 76
Waterbury
    Waterbury Co., 75
Watertown
    Wheeler, Wilson & Co., 75
    Wheeler & Wilson Mfg. Co., 75
West Meriden
    Parkers, Snow, Brooks & Co., 71

DELAWARE

Wilmington
    Charles W. Howland, 76

DISTRICT OF COLUMBIA

Washington
    Post Combination Sewing Machine Co., 73

ILLINOIS

Belleville
    Belleville Mfg. Co., 69
    Thomas M. Cochrane Co., 72
    J.H. Drew & Co., 72
Belvidere
    Eldredge Mfg. Co., 68
    June Mfg. Co., 71
    National Sewing Machine Co., 72
Chicago
    Burnet, Broderick & Co., 67
    Scates, Tryber & Sweetland Mfg., 67
    Chicago Sewing Machine Co., 67
    Sigwalt Sewing Machine Co., 68, 74
    Diamond Sewing Machine Co., 68
    Eldredge Sewing Machine Co., 68
    Foley & Williams Mfg. Co., 70, 73
    Free Sewing Machine Co., 69
    H.B. Goodrich, 70
    Goodrich Sewing Machine Co., 70
    June Mfg. Co., 71
    Parsons Mfg. Co., 74
    Tibbles Mfg. Co., 75
    Wilson Sewing Machine Co., 76
Rockford
    Free Sewing Machine Co., 69
    Illinois Sewing Machine Co., 71
    Royal Sewing Machine Co., 74
Springfield
    Fairbanks Sewing Machine Co., 69

INDIANA

Indianapolis
    Brown Rotary Shuttle Sewing Machine Co., 67

MAINE

Biddeford
    Shaw & Clark Sewing Machine Co., 72
    Shaw & Clark Co., 74

MARYLAND

Baltimore
    Ormund Mfg. Co., 73

MASSACHUSETTS

Boston
    Player, Braunsdorf, & Co., 65
    Braunsdorf, J.E., & Co., 65
    O. Phelps, 66
    J.F. Paul & Co., 66
    Boston Sewing Machine Co., 66
    Bi-Spool Sewing Machine Co., 66
    Acme Mfg. Co., 66
    Daniel Harris, 66
    Bradford & Barber, mfgs., 66
    American Sewing Machine Co., 68
    John P. Bowker, 68
    Empire Co., 69
    M. Finkle, 69
    Finkle & Lyon Sewing Machine Co., 69
    Grover & Baker Sewing Machine Co., 70
    Nichols & Bliss, 70
    J.B. Nichols & Co., 70
    Nichols, Leavitt & Co., 70, 71
    N. Hunt & Co., 70
    Hunt and Webster, 70
    Emery, Houghton & Co., 71
    Ladd, Webster & Co., 71
    Leavitt & Co., 71
    Leavitt Sewing Machine Co., 71
    Safford & Williams Makers, 72
    C.A. French, 73
    F.R. Robinson, 73
    Howard & Davis, 74
    I.M. Singer & Co., 74
    Butterfield & Stevens Mfg. Co., 76
    Williams & Orvis Sewing Machine Co., 76
Chicopee Falls
    Shaw & Clark Co., 74
    Chicopee Sewing Machine Co., 67, 74
Florence
    Florence Sewing Machine Co., 67, 69

Foxboro
    Rotary. Shuttle Sewing Machine Co., 74
    The Foxboro Mfg. Co., 74
Lynn
    Woolridge, Keene and Moore, 67
Orange
    A.F. Johnson & Co., 69
    Gold Medal Sewing Machine Co., 69
    Johnson, Clark & Co., 70, 72, 75
    Grout & White, 72
    New Home Sewing Machine Co., 70, 72
Springfield
    Leader Sewing Machine Co., 71
    Springfield Sewing Machine Co., 74
    D.B. Wesson Sewing Machine Co., 75
Winchendon
    Goodspeed & Wyman, 66
    J.G. Folsom, 69, 72
    William Grout, 72
Worcester
    Goddard, Rice & Co., 66

MICHIGAN

Detroit
    Decker Mfg. Co., 68
    C.R. Gardner, 69

MISSOURI

Saint Louis
    Wardwell Mfg. Co., 75

NEBRASKA

Weeping Water
    The Noble Sewing Machine & Mfg. Co., 73

NEW HAMPSHIRE

Dover
    O.L. Reynolds Mfg. Co., 72
Marlboro
    Thurston Mfg. Co., 67
Mason Village
    Franklin Sewing Machine Co., 69

NEW JERSEY

Bordentown
    Blees Sewing Machine Co., 66

Elizabethport
    Singer Mfg. Co. (manufactory not office), 74
Newark
    Domestic Sewing Machine Co., 68
Paterson
    Whitney Sewing Machine Co., 76

NEW YORK

Binghampton
    Independent Sewing Machine Co., 71
Brooklyn
    Elastic Motion Sewing Machine Co., 68
    J.H. Lester, 71
    G.L. Du Laney, 71
Illion
    E. Remington & Sons, 73
    Remington Sewing Machine Agency, 73
Ithaca
    Aiken and Felthousen (patentees), 65
    American Magnetic Sewing Machine Co., 66
    Clinton Brothers, 67
    T.C. Thompson, 75
New York
    Otis Avery, 66
    Avery Sewing Machine Co., 66
    A. Bartholf, mfg., 66
    Bartholf Sewing Machine Co., 66
    Bartlett Sewing Machine Co., 66
    Barlow & Son, 66
    Beckwith Sewing Machine Co., 66
    Bruen Mfg. Co., 67
    J.A. Davis, 68
    Demorest Mfg. Co., 68
    Charles A. Durgin, 68
    Elliptic Sewing Machine Co., 69
    Empire Sewing Machine Co., 69
    Eureka Shuttle Sewing Machine Co., 69
    Excelsior Sewing Machine Co., 69
    Madame Demorest, 69
    First and Frost, 69
    L. Griswold, 70
    Howe Sewing Machine Co., 70
    Kruse Mfg. Co., 71
    Lever Motion Sewing Machine Co., 71
    Thos. A. Macauley Mfg., 72

    Manhattan Sewing Machine Co., 72
    New York Sewing Machine Co., 73
    T.W. Robertson, 73
    I.M. Singer & Co., 74
    Singer Mfg. Co., 74
    Standard Shuttle Sewing Machine Co., 74
    Henry Stewart & Co., 75
    Stewart Mfg. Co., 75
    United States Sewing Machine Co., 75
    E.E. Lee & Co., 75
    Willcox & Gibbs Sewing Machine Co., 76
    Continental Mfg. Co., 76
Plattsburgh
    Demorest Mfg. Co., 68
    Williams Mfg. Co., 70
Watertown
    Davis Sewing Machine Co., 68
Westmoreland
    A.B. Buell, 67

NORTH CAROLINA

Shelby
    Carolina Mfg. Co., 67

OHIO

Chillicothe
    Melone Sewing Machine Co., 72
Cincinnati
    The Eclipse Sewing Machine Co., 67, 68
    William M. Shaw, 72
    Queen City Sewing Machine Co., 67, 73
    Miles Greenwood & Co., 76
Cleveland
    Wilson [W.G.] Sewing Machine Co., 67, 76
    Daisy Sewing Machine Co., 67
    Domestic Sewing Machine Co., (after 1924), 68
    Leslie Sewing Machine Co., 71
    Standard Sewing Machine Co., 74
    White Sewing Machine Co., 68, 76
Dayton
    Davis Sewing Machine Co., 68
Elyria
    West & Willson Co., 75
Norwalk
    Dauntless Mfg. Co., 68, 73

Wm. A. Mack & Co., and N.S.C. Perkins, 68
Domestic Sewing Machine Co., 68, 70
Springfield
   Royal Sewing Machine Co., 74
   St. John Sewing Machine Co., 75
Toledo
   Jewel Mfg. Co., 71
   Union Mfg. Co., 75

PENNSYLVANIA

Erie
   Noble Sewing Machine Co., 73
Philadelphia
   American Buttonhole, Overseaming and Sewing Machine Co., 65
   American Sewing Machine Co., 65
   Centennial Sewing Machine Co., 67
   George B. Sloat & Co., 67, 68
   Rex & Bockius, 70
   Grant Bros. & Co., 70
   B.W. Lacey & Co., 72
   Parham Sewing Machine Co., 73
   Philadelphia Sewing Machine Co., 73
   Quaker City Sewing Machine Co., 73
   E. Remington & Sons, 73
   Taggart & Farr, 75
   Triune Sewing Machine Co., 75

Pittsburgh
   Love Mfg. Co., 71
Rochester
   Love Mfg. Co., 71

RHODE ISLAND

Providence
   The Brown & Sharp Mfg. Co., 70
   Household Sewing Machine Co., 70
   Providence Tool Co., 70

VERMONT

Brattleboro
   Samuel Barker and Thomas White, 67
   Estey Sewing Machine Co., 69
   Brattleboro Sewing Machine Co., 69
   Higby Sewing Machine Co., 70
   Nettleton & Raymond (Charles Raymond), 72
Windsor
   Lamson, Goodnow & Yale, 67, 76
   Vermont Arms Co., 76

VIRGINIA

Richmond
   Lester Mfg. Co., 71
   Union Sewing Machine Co., 68, 71

Old Dominion Sewing Machine Co., 73

WISCONSIN

Milwaukee
   Whitehill Mfg. Co., 76

LOCATION UNKNOWN

Hibbard, B.S. & Co., 70
Heberling Running Stitch Co., 70
Keystone Sewing Machine Co., 71
Lathrop Combination Sewing Machine Co., 71
Lyon Sewing Machine Co., 71
McKay Sewing Machine Assoc., 72
Moore Sewing Machine Co., 72
C.F. Thompson Co., 75
Finkle and Lyon Mfg. Co., 75
Jones & Lee, 75
Farmer & Gardner Mfg. Co., 75
John W. Beane, 76
Henry Brind, 76
Garfield Sewing Machine Co., 76
Geneva Sewing Machine Co., 76
Gove & Howe, 76
Hood, Batelle & Co., 76
Wells & Haynes, 76

Alphabetical Index to Patentees Listed in Appendix V

Abercrombi, Charles, 520977
Abbott, W. W., 252984
Adams, John Q., 92138
Alexander, Elisa, 16518
Ambler, D. C., 11884
Angell, Benjamin J., 19285
Applegate, John H., and Webb, Charles B., 177784
Appleton, C. J., and Sibley, J. J., 179440
Arnold, B., 86121, 138981, 138982
Arnold, G. B. and A., 30112
Arnold, G. B., 28139
Atwater, B., 44063
Atwood, J. E., 22273
Atwood, K. C., 194759
Atwood, J. E., J. C., and O., 19903
Avery, Otis, 9338, 10880
Avery, O. and Z. W., 22007

Bachelder, J., 6439
Baglin, William, 154113
Baker, G. W., 70152, 125374, 130005
Baldwin, Cyrus W., 38276
Banks, C. M., 225784
Barcellos, D., 203102
Barnes, M. M., 106307
Barnes, William T., 20688, 25084, 25876
Barney, Samuel C., 156119
Barrett, O. D., 25785
Bartholf, Abraham, 19823
Bartlett, Joseph W., 46064, 76385
Bartlett, Joseph W., and Plant, Frederick, 159065
Barton, Kate C., 182096
Bartram, W. B., 54670, 54671, 60669, 62520, 83592, 104247, 130557
Bates, W. G., assignor to Johnson, William H., 9592
Bayley, C. H., 212122
Bean, Benjamin W., 2982
Bean, E. E., 28144
Beardslee, W. F., 245781
Beck, August, 190184, 196863
Beckwith, William G., 126921, 133351, 167382
Behn, Henry, 18071, 18880
Belcher, C. D., 16710
Benedict, C. P., 83596

Bennett, Frank Howard, and Dowling, James, 409728
Bennor, Joseph, 106249
Benson, George, 238556
Beukler, William, 204704
Beuttel, Charles, 115155
Bigelow, J., 263467
Billings, C. E., 88603
Bishop, H. H., 22226
Black, Samuel S., 146642
Blake, Lyman R., 20775, 74289
Blanchard, Helen A., 141987
Bland, Henry, 216016, 221505
Blodgett, S. C., 15469, 21465
Blodgett, John W., 279320
Boecher, Adam, 300199
Bond, Joseph Jr., 12939, 93588, 189599
Bonnaz, A., 83909, 83910
Booth, Ezekiel, 25087
Boppel, Jacob, 396979
Borton, S., 214089
Borton, Stockton, and Willcox, Charles H., 255576, 255577, 255580, 255581
Bosworth, C. F., 19979, 38807, 122555, 199500, 216504
Bouscay, Eloi, Jr., 127145
Bowyer, J. T., 454708
Boyd, A. H., 19171, 24003, 31864
Bracher, T. W., 221508
Bradeen, J. G., 9380
Bradford, E. F. and Pierce, V. R., 177371
Braundbeck, E., 127675
Brewer, A. G., 152894
Briggs, Thomas, 198790
Brown, F. H., 94389, 102366, 131735, 193477
Bruen, J. T., 31208
Bruen, L. B., 68839
Budlong, William G., 25946
Buell, J. S., 25381
Buhr, Johannes, 151272
Burnet, S. S. and Broderick, W., 22160
Burr, Theodore, 32023
Butcher, Joseph, 180542, 233657

Cadwell, Caleb, 45972, 71131
Cajar, Emil, 50299, 61711
Cameron, James W., 272527
Campbell, Duncan H., 241612, 241613
Canfield, F. P., 86057
Carhart, Peter S., 24098

Carlisle, W. S., 270540
Carpenter, Lunan, 20990
Carpenter, Mary P., 112016
Carpenter, William, 87633
Carr, William H. and Ostrom, F. W., 303361
Case, George F., 33029
Cately, William H., 56902
Chamberlain, J. N., 28452
Chandler, Rufus, 133757, 139368
Charmbury, H., 316745
Chase, M., 113498
Chilcott, J. and Scrimgeour, J., 12856
Chittenden, H. H., 43289
Choquette, A. E., 244033
Clark, David W., 19015, 19072, 19129, 19409, 19732, 20481, 21322, 23823
Clark, Edwin, 26336
Clark, Edwin E., 74751
Clements, James M., 57451
Cleminshaw, S., 128363, 213391
Clever, P. J., 96886
Clever, Peter J., 296529
Coates, F. S., 19684
Cole, W. H., 79447
Conant, J. S., 12233
Conklin, N. A., 206774
Cook, Hugo, 454610
Cooney, W., 102226
Corbett, E. and Harlow, C. F., 192568
Corey, J. W., 198970
Corliss, George H., 3389
Cornely, E., 73696
Cornely, Emile, 219225
Craige, E. H., 62186, 67635
Crane, Thomas, 150532
Crosby, C. O., 50225, 90507
Crosby, C. O. and Kellogg, H., 37033
Cummins, William G., 187822
Curtis, G. H. W., 228985
Cushman, C. S., 142442, 184594

Dale, John D., 44686
Dancel, Christian, 199802
Darling and Darling, 163639
Davis, Job A., 27208, 58614
Derocquigny, A. C. F., Gance, D., and Hanzo, L., 34748
Deschamps, O. L., 253915
Destouy, Auguste, 56729
Dewees, John W., 432746

Dickinson, C. W., 26346
Diehl, Philipp, 339623, 347776, 347777, 348113
Dieterle, H. E., 353542
Dimmock, Martial and Rixford, Nathan, 19135
Dimond, George H., 196198, 207400
Dinsmore, Alfred S., 160512, 231155
Dinsmore, A. S., and Carter, John T., 152618
Doll, Arnold, 68420
Dopp, H. W., 27279
Dowling, James, and Connolly, John, 299118
Drake, Ellis, 155932
Duchemin, William, 59715, 91101, 135032, 288929
Dunbar, C. F., 88282
Durgin, Charles A., 12902

Earle, T., 31156
Eickemeyer, Rudolf, 52698, 182182, 239319
Elderfield, F. D., 204429
Eldridge, G. W., 87331
Elliott, F., 85918
Emerson, John, 50989
Emswiler, J. B., 25002
Esty, William, 187837

Fales, J. F., 74328
Fanning, John, 72829, 129013
Farr, Chester N., 25004
Farrar, Arthur, 309837
Fetter, George, 18793, 19059
Fish, Warren L., 123625
Fletcher, James N., 443756
Follett, Joseph L., 189446
Fosket, William A., and Savage, Elliot, 22719, 25963
French, Stephen, 80345
Frese, B., 172308
Fuller, H. W., 63033
Fuller, William M., 32496

Garland, H. P., 159812, 169163
Gee, W. V., 361406
Gibbs, James E. A., 16234, 16434, 17427, 21129, 21751, 27214, 28851
Gird, E. D., 87559
Goodes, E. A., 136718, 147387
Goodspeed, G. N., 54816
Goodwin, Julius C., 450793
Goodwyn, H. H., 24455
Goodyear, Charles Jr., 111197, 116947
Gordon, James, and Kinert, William, 125807

Gray, Joshua, 16566, 19665, 24022, 95581
Green, George F., 238678
Greenough, John J., 2466
Gritzner, M. C., 44720, 76323
Gritzner, Max C., 243444
Grote, F. W., 38447
Grout, William, 24629
Grover, William O., 14956, 33778, 36405, 100139
Grover, William O., and Baker, William E., 7931
Guinness, William S., 41916
Gullrandsen, P. E., and Rettinger, J. C., 180225
Gutmann, Julius, 90528

Halbert, A. W., 76076
Hall, William, 24870
Hale, William S., 36084
Hall, Luther, 43404
Hall, L., 105329
Hall, John S., 168637, 187006
Hallenbeck, J. P., 425422
Hallett, H. H., 191584
Hamm, E., 219578
Hancock, Henry J., 83492, 112033
Hanlon, John, 52847
Happe, J., and Newman, W., 130715, 137199
Hardie, J. W., 30854
Harris, Daniel, 17508, 17571
Harris, David, 185228
Harrison, J. Jr., 10763, 13616, 25013, 25262
Hart, William, 50469
Hayes, James, 55029
Hayden, H. W., 24937
Heberling, J., 204604, 227249, 227525, 284300
Hecht, A., 50473
Heery, Luke, 94740
Heidenthal, William, 127765
Helwig, Arthur, 350364
Hendrick, Joseph E., 19660, 21722
Hendrickson, E. M., 34330
Henriksen, H. P., 104590, 188515, 215615
Hensel, George, 24737
Herron, A. C., 20557
Hesse, Joseph, 235085
Heyer, W. D., 40622
Heyer, Frederick, 30731
Hicks, William C., 26035, 29268
Hinkley, Jonas, 25231
Hinds, Jesse L., 131166

Hine, Charles M., 420382
Hine, Charlie M., 246136
Hodgkins, Christopher, 9365, 33085, 69666
Hoefler, J., 246883
Hoffman, George W., 94112
Hoffman, Clara P., and Meyers, Nicholas, 207035
Holden, O. J., and Griswold, L., 272050
Hollowell, J. G., 27624
Holly, Birdsill, 28176
Hook, Albert H., 22179, 24027
Horr, Addison D., 149862
House, James A., 206239
House, James A., and House, Henry A., 36932, 39442, 39445, 55865, 87338, 114294
Howard, C. W., 126056, 126057
Howard, T. S. L., 199206
Howard, E. L., 154485
Howard, E., and Jackson, W. H., 103745
Howe, Amasa Bemis, 37913
Howe, Elias, 4750, 16436
Hubbard, George W., 18904, 21537
Hull, E. H., 254217
Humphrey, D. W. G., 36617, 49627
Hunt, Walter, 11161
Huntington, Thomas S., 158214
Hurtu, Auguste J., 258761, 315037
Hurtu, Auguste J., and Hautin, Victor J., 98064

Ingalls, N., Jr., 212602

Jackson, Francis, 519064
Jackson, William, 181941
Jacob, Frederick, 190047
Jencks, G. L., 74694
Jennings, L., 16237
Johnson, Albert F., 16387, 20686, 26948
Johnson, W. H., 10597
Jones, John T., 86163, 117640, 169106
Jones, Samuel H., 140631
Jones, Solomon, 118537, 118538
Jones, William, and Haughian, P., 32297
Juengst, George, 27132, 228820
Junker, Carl, 217112

Kallmeyer, G., 137689
Keats, Alphonso, 260990
Keats, John, 198120
Keats, John; Greenwood, Arthur; and Keats, Alphonso, 171622

Keats, John, and Clark, Wm. S., 50995
Keith, Jeremiah, 97518, 170741, 196809, 209126, 243710
Keith, T. H., 196909
Keith, T. K., 170955
Kelsey, D., 24939
Kendall, George F., 101887
Kern, Ferdinand, 479369
Kilbourn, E. E., 59746
Kjalman, H. N., 235783
Knoch, Charles F., 183400
Koch, William, 262298
Koch, Friederich, and Brass, Robert, 138898, 145215

Lamb, Isaac W., 109632
Lamb, Thomas, 98390, 118728
Lamb, Thomas, and Allen, John, 49421
Lamson, Henry P., 79579
Landfear, William R., 109427, 155193
Langdon, Leander W., 13727, 39256
Lathbury, E. T., 17744
Lathrop, Lebbeus W., 139067
Lathrop, Lebbeus W., and de Sanno, William P., 40446
Lawrence, G. H., 290895
Lazelle, W. H., 18915
Leavitte, Albert, 171147
Leavitt, Rufus, 30634
Leavitt, Albert, and Drew, Henry L., 187874
Legat, D. M., 218388
Legg, Albert, and Weston, Charles W., 474840
Leslie, A. M., 241808
Leyden, Austin, 57157
Lingley, John W., 368538
Lipe, C. E., 229322
Lisle, Myron C., 428171
Ludeke, Waldemar, 275506, 300380
Lyon, Lucius, 89489, 96713, 105820
Lyon, W., 12066

Macauley, F. A., 195939
Mack, William A., 38592
Mallary, G. H., 31897
Mann, Charles, 33556
Marble, F. E., 33439
Marin, Charles, 179709
Martin, W., Jr.; Dawson, D. R.; and Orchar, R., 206743
Martine, Charles F., 104612
Meyers, Nicholas, 99783
Melhuish, R. M., 194610
McCloskey, John, 55688, 161534
McClure, A. T., 130385

McKay, Gordon, and Blake, Lyman, 42916
McKay, Gordon, 188809
McLean, J. N., 88499
McCombs, George F., 208407
McCurdy, James S., 24395, 26234, 28097, 28993, 36256, 38931, 46303, 53743
Miehling, Charles, 351992
Miller, Charles, 9139, 26462
Miller, Westley, 20763
Miller, Lebbens B., and Diehl, Philipp, 229629, 274359
Mills, Daniel, 96944, 265850, 313359
Mooney, John H., 222298, 244470
Moore, Charles, 21015
Moreau, Eugene, 110669, 156171
Morley, J. H., 228918, 236350
Morrell, R. W.; Parkinson, T.; Parkinson, J., 202857
Morrison, T. W., 216289
Muegge, C. A., 332207
Mueller, H., 28996
Muir, William, 147152
Murphy, E., 176880

Nasch, Isidor, 104630
Necker, Carl, 117101
Nettleton, William H., and Raymond, Charles, 17049, 18350
Newell, George F., 242470
Newlove, Thomas, 27761
Norton, B. F., 32782

O'Niel, John, 137618
Oram, Henry, 185952
Osborne, J. H., 224219
Otis, S. L., 221093
Osterhout, James A., and Hallenbeck, Joseph P., 402610

Page, Charles, 96343, 150479
Paine, A. R., 27412
Palmateer, William A., 187479
Palmer, Aaron, 35252
Palmer, C. H., 38450, 124694, 240758
Palmer, Frank L., 185954
Palmer, John H., 298228
Parham, Charles, 109443, 135579
Parker, Sidney, 19662, 24780
Parks, Volney, 129981
Parmenter, Charles O., 212495
Payne, R. S., 30641
Pearson, M. H., 199991
Pearson, William, 26201
Pearson, William, 166805, 172478
Perry, James, 22148

Piper, D. B., 56990
Pipo, John A., 37550
Pitt, James; Joseph; Edward; and William, 117203
Planer, Louis, 43927
Porter, D. A., 144864
Porter, Alonzo, 99704
Porter, D'Arcy, and Baker, George W. 174703
Powell, Thomas, 442695
Pratt, Samuel F., 16745, 22240

Ragan, Daniel, 137321
Ramsden, John W., 262116
Randel, William, 192008
Rayer, William A., and Lincoln, William S., 108827
Raymond, Charles, 19612, 22220, 32785, 32925
Reed, T. K., 60241, 62287
Rehfuss, George, 40311, 51086, 61102, 73119
Reynolds, O. S., 19793
Rice, T. M., 176686
Rice, Quartus, 31429
Richards, Jean E., 252799
Richardson, E. F., 145687
Richardson, Everett P., 146948, 165506
Roberts, William, 254696
Robertson, T. J. W., 12015
Robinson, Charles E., 110790
Robinson, Frederick R., 7824
Rodier, Peter, 59659
Roper, S. H., 11531, 16026, 18522
Rose, Israel M., 28814, 31628
Rose, Reubin M., 170596
Rosenthal, Sally Adolph, 353970
Ross, Noble G., 31829
Ross, J. G., and Miller, T. L., 138764
Rowe, James, 26638, 27260
Ruddick, H., 28538
Rudolph, Bruno, 99481
Rudolph, Ernst B. (deceased), Boulter, W. E. (administrator), 449927
Russell, W. W., 86695

Sage, William, 17717
Sangster, Amos W., 21929
Savage, Elliott, 19876
Sawyer, Irvin P., and Alsop, T., 25918
Sawyer, Sylvanus, and Esty, William, 174159
Schmidt, Albert E., 162697
Schwalback, M., 56805
Scofield, Charles, 26059

Scofield, Charles, and Rice, Clarke, 28610
Scribner, Benjamin, Jr., 146483
Secor, J. B., 253772
Sedmihradsky, A. J., 196486
Shaw, E., 230580
Shaw, Philander, 11680
Shaw, A. B., 37202
Shaw, Henry L., 32007
Shaw and Clark, 38246
Sheffield, G. V., 135047
Shorey, Samuel W., 148765
Sibley, J. J., 42117
Sidenberg, William, 112745, 115117
Simmons, Frederick, 216902
Simmons, A. G., and Scofield, C., 41790
Singer, Isaac M., 8294, 10975, 12364, 13065, 13661, 13662, 14475, 60433, 61270
Smalley, J., 27577
Smith, DeWitt C., 45528
Smith, E. H., 20739, 21089, 96160
Smith, H. B., 12247
Smith, J. M., 31334
Smith, Lewis H., 31411, 32385
Smith, John C., 34988
Smith, James H., 148902
Smith, Wilson H., 28785
Smith, W. M., 225199
Smith, William T., 99743
Smyth, D. M., 126845, 234732
Snediker, J. F., 222089
Snyder, Watson, 22987
Spencer, James C., 24061
Spencer, James H., and Lamb, Thomas, 22137
Speirs, John, 152813
Spoehr, F., 101779
Springer, William A., 128919, 142290, 147441
Stackpole, G., and Applegate, J. H., 220314
Stannard, M., 64184
Stedman, G. W., 12074, 12573, 12798, 13856
Stein, M. J., 81956, 113593
Stevens, George, and Hendy, Joshua, 111488
Steward, A., 207454

Stewart, James, Jr., 141397, 432449
Stewart, W. T., 205698
Stoakes, J. W., 32456
Sullivan, John J., 179232
Sutton, William A., 29202
Swartwout, H. L., 89357
Swingle, A., 14207

Taft, J. C., 547866
Tapley, G. S., 25059
Tarbox, John N., 49803
Tate, William J., 113704
Taylor, F. B., 146721
Temple, John, 358088
Thayer, Augustus, 172205
Thimonnier, E., and Vernaz, C., 287592
Thomas, Joseph, 330170
Thomas, J., 236466
Thompson, J., 27082
Thompson, T. C., 9641
Thompson, Rosewell, 34926, 42449
Thurston, C. H., 233300
Tittman, Alexander, 89093, 136792
Toll, Charles F., 171193
Tracy, Dwight, 30012
Tracy, Dwight, and Hobbs, George, 40000
Trip, J., 308711
True, Cyrus B., 148336
Tucker, R. D., 320898
Tucker, Joseph C., 56641
Turner, M. G., 312306
Turner, S. S., 133553
Tuttle, J. W., and Keith, T. K., 219782
Thompson, George, 115255

Uhlinger, W. P., 21224
Upson, L. A., 176153

Van Vechten, O. R., 302063
Varicas, L., 204864
Venner, O., 133814
Veukler, W., 255916

Wagener, Jeptha A., 40296
Walker, William, 141407, 176101, 320099, 353720
Walker and Bennet, 428548
Ward, D. T., 12146
Wardwell, Simon W., Jr., 128684, 141245, 148339

Warth, Albin, 56646, 73064
Washburn, T. S., 30031
Waterbury, Enos, 79037
Watson, William C., 14433, 18834
Weber, Theodore A., 145823, 166236
Webb, T., and Heartfield, C. H., 213537
Webster, W., 182249, 402497
Weitling, W., 37931, 45777
Wells, W. W., 209843
Wensley, James, 152055, 207230
West, Elliott P., 117708, 130674, 138772
West, H. B., and Willson, H. F., 20753
Wheeler, Nathaniel, 420847
Wheeler, Nathaniel, and Dial, Wilbur F., 328165
Wheeler, Darius, and Carpenter, Lunan, 21100
Whitehill, Robert, 166172
Wickersham, William, 9679, 18068, 18069
Wilder, M. G., 32323
Wilkins, J. N., 36591
Wilkinson, Charles E., 269251
Willcox, Charles H., 42036, 43819, 44490, 44491, 218413, 230212, 246700
Willcox, C. H., and Borton, S., 472094
Willcox, C. H., and Carleton, C. 116521, 116523, 116783
Wilson, Allen B., 7776, 8296, 9041
Wiseman, Edmund, 228711
Witherspoon, S. A., 176211
Winter, William, 88936
Wollenberg, H., and Priesner, J., 206848
Wood, John, 185811
Wood, Richard G., 207928
Woodruff, George B., and Browning, George, 97014
Woodward, E., 247285
Woodward, F. G., 25782
Woodward, Erastus, and Keith, Thomas K., 316927
Wormald, William, and Dobson, Edmund, 169881

Young, E. S., and Dimond, G. H 206992

General Index to Chapters 1-4

Adams and Dodge, 9
Aetna Sewing Machine Company, 40
American Buttonhole and Sewing Machine Company, 40
Archbold, Thomas, 13
Arrowsmith, George A., 11

Bachelder, John, 22, 30, 34, 41, 42
Baker, William E., 36 (*see* Grover & Baker)
Bartholf, A., 24, 40
Bartlett Sewing Machine Company, 40
Bartram & Fanton Manufacturing Company, 40
Bean, Benjamin W., 13, 14, 15
Blees Sewing Machine Company, 40
Bliss, George, 24
Blodgett and Lerow, 24, 26, 30
Blodgett, Sherbrune C., 25
Bradshaw, John A., 21, 22, 26, 27
Brown, W. N., 50

Centennial Sewing Machine Company, 40
Chapman, Edward Walter, 7, 19
Chapman, William, 7
Clark, D. W., 47, 49
Clark, Edward, 33, 34, 35
Combination, Sewing-Machine, 23, 24, 38, 41–42, 47, 48
Conant, Jotham S., 22
Corliss, George H., 14, 15, 16

Dale, John D., 54
Davis, Ari, 19
Davis Sewing Machine Company, 40
Demorest, Madame, 53
Dodge, Rev. John Adam, 9
Domestic Sewing Machine Company, 40
Duncan, John, 6, 19

Elliptic Sewing Machine Company, 40
Ellithorp, S. B., 51
Ellithorp & Fox, 51
Empire Sewing Machine Company, 40

Fairy Sewing Machine, 53
Family Sewing Machine (Singer), 35
Finkle & Lyon Manufacturing Company, 40
Fisher, George, 19
Fisher, John, 15, 16
Florence Sewing Machine Company, 40
Folsom, J. G., 40

Gibbons, James, 15
Gibbs, James E. A., 45, 48
Goddard, Rice & Co., 25
Gold Medal Sewing Machine Company, 40, 53
Goodspeed & Wyman Sewing Machine Company, 40
Grasshopper, The, 35
Greenough, John J., 13, 14
Grover & Baker Sewing Machine Company, 24, 35, 37, 38, 40, 41
Grover, William O., 35, 38

Heberling, John, 54
Henderson, James, 6
Hendrick, Joseph, 49
Heyer, W. D., 52
Hook, Albert H., 50
Howe, Amasa B., 24
Howe, Elias, Jr., 11, 18, 19, 20, 21, 22, 23, 24, 28, 29, 33, 34, 41, 42, 138 (biographical sketch)
Howe Machine Company (Elias), 25
Howe Sewing Machine Company (Amasa, then B. P.), 24, 25, 40
Hunt, Walter, 10, 11, 19, 33, 138 (biographical sketch)

Jenny Lind (sewing machine), 30
Johnson, Joseph B., 22

Keystone Sewing Machine Company, 40
Kline, A. P., 27
Knowles, John, 9
Krems, Balthasar, 7, 19

Ladd & Webster Sewing Machine Company, 40
Leavitt Sewing Machine Company, 40
Lee, Edward, 27
Lee, E. & Co., 27, 28
Lerow, John A., 24, 25
Little Gem, 53
London Sewing Machine, 22
Lye, Henry, 9

McKay Sewing Machine Association, 40
Madersperger, Josef, 8, 9, 12, 13
Magnin, Jean Marie, 11, 22
Mason, The Honorable Charles, 13
Morey, Charles, 22
Morey & Johnson, 23, 34, 42

Newton, Edward, 13
Nichols and Bliss, 24

Palmer, Aaron, 52
Parham Sewing Machine Company, 40
Perry, James, 49
Phelps, Orson C., 25, 30, 31
Potter, Orlando B., 37, 38, 41

Remington Sewing Machine Company, 40
Robertson, T. J. W., 47
Rodgers, James, 15

Safford & Williams Makers, 22
Saint, Thomas, 4, 5, 19
Secor Sewing Machine Company, 40
Shaw & Clark Sewing Machine Co., 40, 54
Singer, Isaac Merritt, 23, 25, 30, 31, 33, 34, 35, 41, 142 (biographical sketch)
Singer, I. M. Company, 13, 32, 34, 40
Singer Manufacturing Co., 30, 40, 42
Singer's Perpendicular Action Sewing Machine, 30, 31
Stone, Thomas, 6

237

Thimonnier, Barthelemy, 11, 19, 137 (biographical sketch)
Thomas, William, 20
Thompson, C. F., 40
Turtleback Machine (Singer), 35

Union Buttonhole Machine Company, 40

Warren and Woodruff, 28
Weatherill, Jacob, 37
Weed Sewing Machine Company, 40
Weisenthal, Charles F., 4, 19
Wheeler & Wilson Manufacturing Company, 29, 30, 40
Wheeler, Wilson, Company, 24, 28, 41
Wheeler, Nathaniel, 27, 28
Willcox & Gibbs Sewing Machine Company, 40, 46, 48
Willcox, Charles, 46, 48
Willcox, James, 46
Wilson, Allen Benjamin. 26, 27, 28, 29, 30, 141 (biographical sketch). See Wheeler & Wilson.)
Wilson, Newton, 4
Wilson, (W. G.), Sewing Machine Company, 40
Woolridge, Keene, and Moore, 24

Zieber, George, 30, 33